Zero Outage

Stephan Kasulke • Jasmin Bensch

Zero Outage

Putting ICT Quality First
in the Digital Era

Opening Remarks by Ferri Abolhassan

 Springer

Stephan Kasulke
T-Systems International GmbH
Vienna, Austria

Jasmin Bensch
T-Systems International GmbH
Bonn, Germany

ISSN 2192-8096   ISSN 2192-810X (electronic)
Lecture Notes in Information Systems and Organisation
ISBN 978-3-319-53738-2   ISBN 978-3-319-53739-9 (eBook)
DOI 10.1007/978-3-319-53739-9

Printed on acid-free paper

This Springer imprint is published by Springer Nature
The registered company is Springer International Publishing AG Switzerland

The Authors

Stephan Kasulke has been Senior Vice President of Quality at T-Systems International GmbH since 2012. In this role, he manages the worldwide Zero Outage program, which brings improvements to the quality of operations and projects. Within its first three years this program reduced the quantity of moderate and severe operational incidents by 95 percent, whilst customer satisfaction scores increased from average – as compared to competitors – to market leadership levels.

Kasulke began his career in 1985 at the age of 16 as an independent software developer for Weidmüller and other regional companies in East Westphalia in Germany. Beginning in 1995, he developed sales systems for Deutsche Leasing AG in Bad Homburg; in 2001 he was appointed as Head of IT Infrastructure and Architecture. In 2004, Kasulke joined GE Capital where he held several management positions. In 2007 he was appointed CIO of GE Money Germany. From 2009 until 2012, he was Managing Director of IT Solutions in Vienna, Austria, a company that develops and operates ICT solutions and applications for eight banks in Central and Eastern Europe belonging to Erste Group Bank AG.

Kasulke holds an MBA from the University of Maryland, USA.

Jasmin Bensch has been Executive Consultant ITIL and Head of Line Office Quality at T-Systems International GmbH since 2012. She has been instrumental in the strategic refinement, cultural evolution and internal and external communications of the Zero Outage program.

Bensch started her career in 1999 with a dual on-the-job training and study program in information systems in the IT division of Aral AG, later Deutsche BP AG, in Bochum, Germany. Her focus was on SAP development and programming in the field of finance and controlling.

After completing her studies in 2004, Bensch joined RWE IT GmbH, where she held several positions in infrastructure development and served as Staff Director for executive management. In 2008, she became responsible for the international harmonization of change processes, as well as release, configuration and problem management. She was also the quality manager responsible for customer service.

In 2010, Bensch received an MBA from FOM University of Applied Sciences in Essen, Germany, in cooperation with Pfeiffer University, Charlotte, North Carolina, USA.

Foreword

Losses of 2.2 million dollars per hour. This was the estimated damage suffered by an app store when technical problems took out its services for up to 11 hours in 2015 (see ZDNet 2015). Incidents like this are common. In 2013, a major search engine was calculated to have lost hundreds of thousands of euros due to an outage lasting just a few minutes. And in the following year, a system crash at a bank threatened transactions worth billions. Outages happen for many different reasons, and technology alone is not always to blame. Elements of human error and unpracticed process chains can often endanger the availability of services and the security of processes. This occurs in all industries and sectors, from food production to IT departments, operating rooms and self-driving cars. Outages, therefore, not only result in economic losses and damaged reputation; depending on the situation, the risks could be much worse.

To put it plainly: the security and quality of workflows, processes and products are vital – and they should be a top priority for company managers. As a pioneering ICT and cloud provider, we thus faced a great responsibility right from the start: ensuring that our customers could continue to do business, that their products were produced in time, and that our customers' customers were satisfied. Although providing IT and telecommunications services may sound simple, it was and continues to be essential. As an ICT provider – regardless of whether you are an external partner or internal department – you are ultimately responsible for the success and business capabilities of a company.

And in this role, you cannot rely on technologies alone. Long-term quality assurance is equally, if not, more important. It's very simple: If the quality isn't right, the subsequent product won't be right either. I experienced my personal "Aha!" experience in 2010. We had grown very quickly at the time and won a number of large contracts. This meant that we faced several Herculean tasks and had to handle various megaprojects all at once. As a result, the demands placed on us also increased tremendously. However, we wanted to set ourselves up for the future so that, despite the growing complexity, we could maintain the high-quality standards we had always aspired to – for the benefit of our customers. After all, there is nothing less at stake than the long-term business success of our corporate customers.

Therefore, we focused on our processes and started on a blank slate. We put every process to the test. Were there hidden errors anywhere? How could we permanently meet our high-quality demands? Our first, very decisive step was something that we have since institutionalized: we started an intensive and somewhat ruthless dialogue with our customers. This was accompanied by an extensive internal cause analysis. What we discovered is that only a holistic strategy will lead to the desired result. Because optimization measures in just a few obvious areas aren't enough to offer customers **end-to-end** quality.

Zero Errors, Total Engagement

As our analysis clearly showed, it was important for us to define what had to be done when and how. We knew that if we wanted to convince our customers in the long term, we had to define standards and permanently maintain them. In essence, this was the birth of **Zero Outage**.

What exactly did we want to achieve with the program? Our goal was to reduce the number of IT outages to zero – and make the quality of our services truly measurable. After all, you can only improve what you can measure. Thanks to our ruthless error analysis, we also knew what we needed to achieve this goal. Firstly: optimally trained employees who receive regular instruction and certification – with a unified culture of discipline and precision. Secondly: simple, standardized processes, enabling greater efficiency and a high-quality result – with as little implementation risk as possible. And thirdly: standardized, high-performance and high-availability platforms that are always state of the art. We know from our own experience and analyses of other companies that standards are generally lacking in these three core areas: people, processes and platforms.

How Does Zero Outage Work?

What is the best way to tackle such a comprehensive program? After all, you have to walk before you run. In the quick-fix phase, we initially concentrated on the biggest problems of our top 25 customers. To do this, we analyzed all of our customers' business-critical systems and documented them in a **critical landscape**. How important was the ICT system to the customer's business success? And how vulnerable was the respective system? By consistently posing and answering questions such as these, we were able to prioritize and solve existing quality problems based on the respective **customer business impact.** That is, if an ICT system disruption would have serious consequences for a customer's business success, we took care of this issue first. This made it possible for us to quickly improve our level of quality.

Risk categorization was another item on our agenda. We did not want to solve problems in a solely reactive way. Instead, with Zero Outage, we wanted to prevent problems from arising in the first place. We know that in IT and telecommunications – as in many other sectors – there is no such thing as 100-percent stability. But our goal was to get as close to perfection as possible. The **Quality Roadmap** that we devised assisted us here. In this roadmap we recorded 280 individual risks in 40 categories. These included risks such as electricity outages, defective components and employee strikes. We then thought about the concrete measures we could take to prevent such problems. This made it possible for us to proactively manage risks that threatened the reliability of our services and thus, the IT and telecommunications of our customers.

Quality Is a Management Issue – Day and Night

But even with extensive and careful preventive measures, IT disruptions cannot be entirely avoided. So, what happens when an incident does occur? You have to respond immediately and in a deliberate manner to get your customer's systems up and running again as swiftly as possible. This is why we established the **Manager-on-Duty service** as an integral part of the program. The Manager on Duty and respective teams are available worldwide around the clock, 24 hours a day, 365 days a year. And when an incident occurs, they work on the problem until the systems are up-and-running again.

One key aspect of this is that a senior manager is always involved; so, decisions can be made quickly at the highest level. To ensure that everything works seamlessly in the event of an incident, we simulate 500 incidents every year. In these so-called "fire drills," our employees practice every process so that in real emergency they won't put a foot wrong.

Interplay between Humans and Technology

The human factor plays an important role in quality. It is employees who bring quality to life and actually implement it day after day. Another key element of our Zero Outage program has always been our internal **Quality Academy**, a training program for employees that has now expanded to include partners as well. More than 20,000 employees and nearly 100 top partners and access providers have since been certified. This ensures a shared understanding of quality and solution know-how at every level.

These solutions must naturally also meet the most stringent technical requirements. Redundant data center technologies are therefore the infrastructure foundation of Zero Outage. The principle here is that all data and systems are maintained in two architecturally identical, but physically separate data centers. If one data center experiences an outage – due to force majeure, for example – its "twin" immediately jumps into action. This allows us to offer our customers IT availability rates of up to 99.999 percent – which equates to one potential outage of just a few minutes every year. Furthermore, the latest security solutions protect our systems and information from unauthorized access by third parties. Finally, the issue of security must go hand in hand with high standards of quality. Quality and security together is the prerequisite for all **cloud** and digitalization endeavors.

To sum up: redundant, highly secure technologies, clearly defined processes, and qualified personnel are the foundation of Zero Outage. Complemented by preventive measures such as the Manager-on-Duty service, the Zero Outage program ensures the highest possible quality and availability of our ICT services. This type of program is probably unique in the industry.

Measurable Success – Multiple Customer Benefits

The success of this concept can be seen in the results of the Zero Outage program at T-Systems. Thanks to Zero Outage, we have minimized the quantity of major incidents. Each year we carry out around 315,000 IT changes – with a success rate of 99.5 percent. And this is reflected in our customer satisfaction rates. We have achieved the best scores three years in a row, making us a benchmark for the industry. External experts have also confirmed the Zero Outage principle: in 2015, we received the seal of approval from TÜV Rheinland for our quality program.

And every single day we work to improve even more. After all, the Zero Outage approach is of fundamental strategic importance for us. It is the basis of our quality leadership – and it distinguishes us in the market. It is also the foundation for new solutions such as the "Un-outsourcer." We can offer **outsourcing** without long contractual commitments and with cancellation rights in the event of dissatisfaction, because we are completely confident in our high quality and reliability – and this, too, is thanks to Zero Outage.

Setting Standards with Zero Outage

Zero Outage is the central lever for quality on all levels – and it is a standard that can be applied to every other industry and company. This is why we want to share our experiences with you in this book and explain how Zero Outage works in practice. The articles here should help guide you and pave the way for making your own (ICT) processes even more reliable and (fail-)safe. In the following chapters, we share our key findings in quality management and give you an exclusive insight into Zero Outage.

Enjoy the book!
Ferri Abolhassan,
Director Service Transformation Telekom Deutschland

Dr. Ferri Abolhassan

After receiving his doctorate in computer science, Abolhassan began his professional career in R&D at Siemens in Munich, Germany, followed by several years at IBM in San Jose, USA. In 1992, he joined software vendor SAP, where he held a number of senior positions until 2001, including a spell as Senior Vice President of the global Retail Solutions business unit. Following a 4-year tenure as Co-CEO and Co-Chairman at IDS Scheer, he returned to SAP in 2005, most recently as Executive Vice President, Large Enterprise for EMEA. In 2008, Abolhassan took over the newly-created position of Head of Systems Integration at T-Systems and at the same time joined the company's Board of Management. His management portfolio was later expanded to include the Production unit. In 2013, Abolhassan was appointed Director of Delivery before becoming Director of the IT division in 2015, overseeing approximately 30,000 employees and 6,000 customers. In December 2015, Abolhassan also took over the task of setting up the Telekom Security business unit, which will bundle all the security departments in the Deutsche Telekom corporation. In October 2016, he moved to Telekom Deutschland, where he heads the newly created business area Service Transformation as its managing director.

In 2011, he launched the successful "Zero Outage" program to safeguard T-System's quality standards in the face of growing process complexity. Twenty-one thousand employees and key service providers were trained, countless processes standardized and high-availability platforms established. This has resulted in a continuous success rate of 99.5 percent based on 315,000 IT changes annually. Not only has the program been certified by Germany's technical inspection agency TÜV, but customer satisfaction has also risen to its highest level in the company's history, setting a new benchmark in the industry. In 2016, T-Systems worked with partners to create a new industry standard based on the Zero Outage principle.

Table of contents

Introduction – Zero Outage as a Guiding Principle for IT

"Cheap is cool!" trumpeted a well-known German electrical goods retailer in 2002. And this saying has become rather more than a fashionable advertising jingle: it has become anchored in the minds of many people, including decision-makers. Its presence is felt in both our personal and working lives, and since 2007 the global financial crisis has amplified the concept. Today, throughout the economy and year after year, we believe that prices for products and services must fall, and that the same profit must be earned with less costs than in the previous year, or – ideally – profit must go up while costs go down.

For IT decision-makers – whether as a supplier or within an IT department – this model has meant only one thing since the turn of the millennium the best-case forecast for the overall budget is no improvement. This is because cost increases due to higher pay, lease prices, energy costs or higher fees for licenses and maintenance work must be balanced by greater productivity. Business departments and consumers alike have come to assume that enhanced functionality, more resources (in terms of storage or processing power, for example) or improved usability can be had at last year's prices – or lower.

But a perfect storm is brewing, since the **criticality** of IT has increased consistently during the past 30 years. Shortcomings in IT quality can have deadly consequences. In many areas of modern life, even a temporary IT systems outage can result in severe financial losses for the affected companies and facilities, and may even create life-threatening situations. Every production line is controlled by IT systems: for example, complex **supply chains** in the automotive industry rely on the exchange of information and data 24 hours a day to enable real-time collaboration across company boundaries and national borders. For key IT systems, a few hours of downtime can bring an entire sector's production chain to a shuddering halt, worldwide. And there is more: all of our telecommunications, news, TV, police systems, rescue services, drilling platforms, parcel logistics, freight forwarders and ocean freight, retail, and the entire financial sector – yes, even the harvesting of sugar beet – are entirely dependent on fully-functional IT systems. These systems must work 24 hours a day, 365 days a year with virtually no downtime. So, while the need for failure-tolerant IT may not always be immediately obvious to everybody, it is still a necessity. In the 21st century, a CIO is reprimanded if costs are too high, but fired if quality defects are too severe.

This isn't news to decision-makers. As a PwC report has shown, a fundamental rethink of the CIO role occurred around 2012 (see PwC 2012). Contrary to what some might believe – and despite considerable financial pressure and a global economic crisis – price was not the most common aspect named as a key selection criterion for IT suppliers (58 percent). Instead, CIOs focused on quality (84 percent) (see Fig. 1.1).

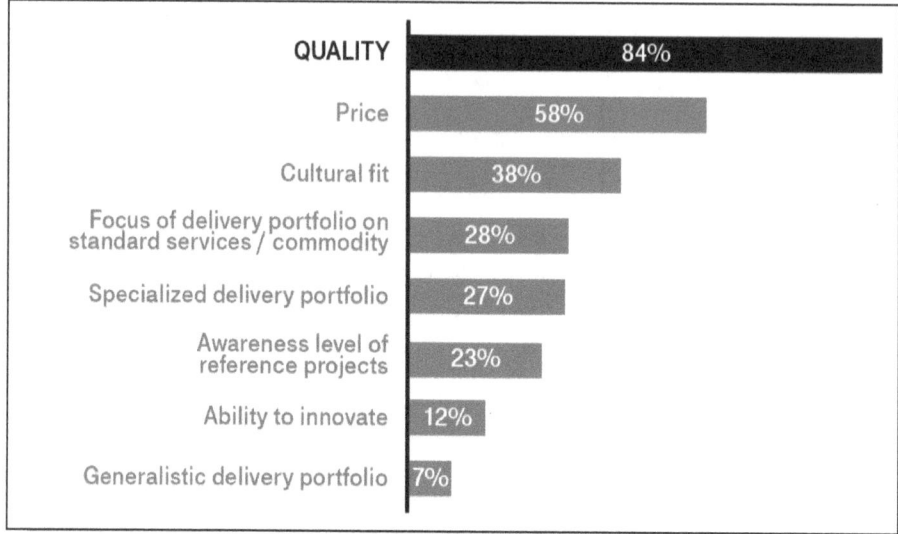

Fig. 1.1 Which Abilities of an ICT Provider Are Important to IT Responsibles?
© PwC Group (slightly adapted figure)

1.1 IT Quality – A Success Factor in the Digital Age

Year after year, IT managers must therefore master the difficult balancing act of increasing the quality they deliver while cutting the costs involved in doing so. At first, this seems an insoluble problem: as already mentioned, budget savings achieved by falling prices for hardware are typically outweighed by an increased demand for resources, higher pay and other costs. The only option, therefore, is to offer improvements in internal processes and organization to deliver better quality despite a shrinking budget. But what exactly does "quality" mean for IT?

The quality of IT services is very much in the eye of the beholder, i.e., the consumer of these services. In a Zero Outage context, three aspects are of fundamental importance here:

1. the stability of the production IT systems currently in use,
2. the reliability of new IT systems in projects and during important delivery, and
3. a visible and professional customer interface.

This book takes both a theoretical and practical look at how systematic application of the Zero Outage concept can be used to measure and – above all – decisively improve these three factors. Interested readers will be shown how a high level of customer satisfaction can be achieved even as costs are cut and demand for resources rises.

1.2   Zero Outage – A Roadmap for the Future

Zero Outage is not a reinvention of a quality system; it is neither a silver bullet nor a buzzword.

Instead, Zero Outage epitomizes the way in which an organization behaves – in terms of the systematic and efficient processing of quality-relevant tasks and assignments aimed at a continuous improvement of customer satisfaction.

Therefore, Zero Outage covers operations in telecommunications and IT, the delivery of service change requests and projects, as well as optimizations to the customer interface and the customer's perception of the ICT supplier. What's also important is that Zero Outage affects the behavior of each and every member of an organization – from senior management downwards. This includes basic behavioral patterns such as:

- a **sense of urgency** – the ability to act quickly, decisively and effectively in critical situations, and with all available resources,
- a readiness to focus on the specifics of a problem,
- accepting the principle of dual review ("**four-eyes principle**") as a prudent strategy for customer-relevant changes and
- a proactive approach to systematic prevention.

Alongside technical principles of redundancy, Zero Outage also involves observing global process standards for operations and projects. And it also implies a strong focus on raising workforce awareness, bolstered by staff training and re-training.

So, Zero Outage is an integrated topic: if quality is to be anchored in a truly sustainable way within the company, it needs to involve every employee at every level of the hierarchy. Quality in the sense of satisfied customers – and in the sense of the company's own profitability. But Zero Outage impacts much more than the company's own business – it has now also a matter of social responsibility. As the impact of faults in communications and IT continues to be felt further afield, customer patience with such faults wears ever more thin, just as the dependency on error-free IT infrastructure continues to rise. In a nutshell, both business and society at large need Zero Outage to ensure that fewer system outages occur around the world, with the resulting improvements in living standards for everyone. A concerted effort must now be made to promote Zero Outage in the enterprise.

2.1   Organizing IT Services: The Zero Outage Principle

As with any local water or electricity supply, IT is now a basic precondition for modern customers, who simply assume that it will be available and working fault-lessly all the time. The choice of providers is no longer the determining factor, since the market offers a comparable set of solutions and services. Accordingly, customers do not select a specific product. Instead, they are looking for maximum value in terms of quality and reliability. For IT service providers trying to attract customers, the ability to offer a high level of quality, even as market prices decline, is therefore an increasingly important differentiator. As competition becomes ever more international and many companies operate branches all over the world, the ability to maintain quality standards across borders also plays a major role. Ide-ally, customers will want to source all of their global IT services from one and the same provider.

As in any relationship, trust is the cornerstone for building long-term collabora-tion between customers and their IT partners. To allow this trust to form and develop, the IT service provider must be in a position to operate the IT systems reliably over an extended period of time. Each fault and outage impacts the trust on which the relationship is built. And this is entirely understandable: these incidents affect com-pany data, sensitive information, the ability of employees to do their work the world over – indeed, the very ability of the company to do business. Apart from scalable and tailor-made solutions, today's customers therefore also expect their IT service providers to offer the greatest possible level of data protection and safeguarding against outages. And the longer the partnership lasts, the more important the aspect of reliability becomes: once customers experience stable operations without outages for a period of time, this then simply becomes the status quo – and the foundation for the customer's business processes. Email – a typically well-functioning system – has become one of the most important components in corporate communications, for example, making sudden outages all the more catastrophic. Should the corporate

Exchange Server unexpectedly go down, communication throughout the company will be massively disrupted.

IT service providers therefore have the ongoing task of improving their **service quality and reliability while continuing to cut production costs** measurably. Industry reports have confirmed this trend: in 2015, for example, the ITSM Group discovered that 70 percent of companies with revenue exceeding 50 million euros considered IT service quality to be a "consistently critical factor for success" – and this perception has been trending upwards since 2007 (see ITSM Group 2015 and Fig. 2.1).

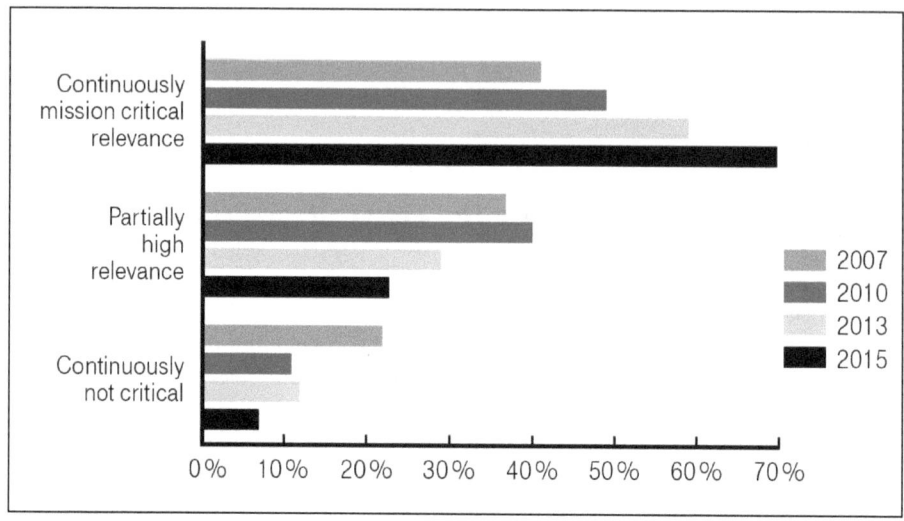

Fig. 2.1 In How Far Are Business Processes Dependant on a High and Measurable IT Service Quality? © ITSM Group (slightly adapted figure)

The importance of IT service quality as a critical success factor stems from a simple fact: outages cost businesses money. A lot of money. In Europe alone, over 37 million working hours are lost due to IT outages and data restoration by companies with over 50 employees. And that's only the annual figure. Beyond this, longer outage periods can threaten the company's very existence if they happen to impact business-critical systems. During the financial crisis, a technical fault meant that smaller banks found themselves in serious trouble. At the time, German investors were extremely concerned about their portfolios, and these worries had been further exacerbated by the Lehman Brothers bankruptcy and the failure of several Icelandic banks. And rightly so: having previously attracted German investors with exceptionally high rates of interest, the banks simply took their online banking offline in 2008, denying account holders access to their money. Understandably, this behavior had made investors extremely skittish – with fatal consequences for other banks. When the online banking service of a U.S. financial service provider went offline for several hours due to a simple technical fault, customers panicked. Thinking they would soon lose access to their money, they made withdrawals totaling several hundred million dollars – a severe shock to the system for the short-term liquidity of a relatively small bank (see Kasulke 2013a).

2.2 Customers: Optimal Service and Continuous Improvement

So, what tools can we use to positively influence the customer's subjective opinion? What are the aspects that build trust in a business relationship? From the customer's point of view, (the subjective perception of) service quality depends on the long-term fulfillment of certain criteria (see Kasulke 2014):

- **Reliability:** Stability of important systems should be optimal and serious outages should never occur. Also, the customer expects the service provider to reliably comply with change requests, guaranteed delivery dates and project costs.
- **Credibility:** Credibility is fostered by clear communication and reporting in a language the customer understands. Even if a fault occurs, IT service providers give points for explicit and – above all – rapid action.
- **Flexibility:** In the event of new or changing requirements, a flexible attitude works wonders to boost customer confidence.
- **Competence:** Compliance with service level agreements (SLAs) is a basic and essential principle. And the IT service provider should not just respond passively to the customer, but also advise proactively on technical developments.
- **Understanding:** IT service providers show understanding by displaying knowledge of the customer's industry and how it is enabled by IT.
- **Security:** Observe the standards necessary to ensure a high level of security and all legal requirements. By employing reliable staff, IT service providers also demonstrate their team-building skills and the efficiency of their business.
- **Contact:** Providing responsive and friendly advice helps the IT service provider to build confidence in the business relationship.

As stated above, this is an area where service quality should measurably increase over time while production costs fall.

2.3 Problems Encountered in Practice

In order to guarantee quality and reliability, it helps to be aware of typical issues, so that these can then be readily identified and resolved. In practice, systematic problems often go unnoticed for a long time, i.e., they are treated as isolated incidents by the supplier and initially tolerated by the customer as "freak accidents" or isolated faults. Only when a spate of problems occurs over a shorter time frame are matters then escalated via the customer executive or CIO, who must now also face the concerted criticism of their own departments. The root causes to such problems are often to be found on both sides – supplier and customer – and can be subdivided into three categories: problems in project planning, problems in the provisioning of operational services, and problems at the interface between the customer and the supplier.

2.3.1 Problems in Project Planning

This challenge is well illustrated by a real example. A few years ago, a company started a major project. The intention was to replace the entire legacy IT backend with a modern and totally new IT environment. For economic reasons, and the fact that the in-house IT unit comprised just under 100 employees, the project was too big to be handled in Germany. The new development was then launched with 300 additional staff members from an Indian software company. At the same time, construction had just started across the road on a new multi-storey office building with underground parking. What did these two projects have in common? One simple fact: the sign providing information about the new construction gave almost the same completion date as was announced for the major IT project. To cut a long story short: while the office building was completed on the stated date, IT management faced the unwelcome task of communicating a delay of several years, together with a new project plan. Of course, people have been building houses for thousands of years – and IT solutions only for a few decades.

Apart from personal experience, many analysts have offered reasons why projects fail, exceed budgets, or are finished with significant delays. Some of the most common causes include:

- **Poor project preparation:** There are many examples of substandard project preparation. A vague definition of what should be delivered, the concrete expectations and requirements of the client, or the acceptance criteria that must be satisfied. On the client side in particular, there is often a lack of understanding about the very real need for client involvement in a project. This leads to numerous change requests during the course of the project, with associated delays and unplanned cost overruns. To make a comparison with our construction project example: if the client were suddenly to decide, half-way through the project, that eight storeys are needed instead of four, this would incur huge additional costs and could even require demolition and re-build. Precise formulation of requirements and objectives is therefore essential before the project starts. In an IT context, however, it is often the case that the client has not properly thought through even the most basic requirements beforehand. Here, the IT service provider must use a structured set of questions to exclude as many errors as possible well in advance and ensure that project preparation is optimal.
- **Poor project management:** Like any other trade, project management consists largely of skills that must be learned. Thanks to the use of standards

such as PMI or PRINCE2 (see Chapter 3), the average level of training possessed by project managers has improved significantly over the last ten years. In complex and large-scale projects, however, hands-on project management must be supplemented by a high degree of professional communication, leadership skills and stakeholder management. And since experienced senior project managers are scarce in most companies, the deployment of less experienced colleagues often leads to errors in project execution.

- **Overly optimistic assumptions:** Typically, important projects start as top-down initiatives led by the company's executive management, with the aim to make change happen. Critical voices inside the company are dismissed as doubters or blockers, and the risks they articulate are swept aside in order to start the project quickly. The result is an overloading of the staff and a failure to manage risks. The project therefore starts with undue optimism and later becomes bogged down in reality.
- **Excessive demands on the team/complexity exceeds ability:** In the example above, the situation quite simply asked too much of the client's company, since none of those involved had ever worked on a project of this magnitude and complexity. New and specific processes had to be implemented, and new tools had to be procured in order to manage the sheer size of the undertaking – plus, the overall planning itself was several degrees more complex than any prior project. The level of professionalism – and thus of reliability – shown in project handling depends on employees' experience.
- **Poor communication:** Projects are dependent on continuous, uninterrupted communication, and stakeholder involvement is crucial, since change is constant. If this communication is disrupted in any way, for example between the client and contractor or between project management and technical implementation, then difficulties will develop quickly. And even where virtual teamwork is going well, it is recommended to locate all of the project team in the same physical space during critical phases, since informal communication can be also a crucial success factor.
- **Changes in the project environment:** If changes in the company management occur at the client, for example, these often result in changes to strategy and prioritization. Also, where projects are set to run for over two years, there is a risk that changes in the client's competitive environment cause changes in project deliverables, which may have a negative effect on delivery and costs.

From a customer perspective, operational incidents are even worse than project delays. What are the typical causes?

2.3.2   Problems in the Provisioning of Operational Services

In almost all branches of industry, a company's most basic processes are dependent on communications and IT. There is no practical way to work around an operational incident in an ERP system with manual methods and processes. Within a short space of time – often a few minutes, at most a few hours – the customer's business will be at a standstill. Company employees will be sent home and customers will be left high and dry. This not only causes large-scale losses but, depending on the outage severity, can also lead to a massive loss of reputation for the customer – and then (if not earlier) an escalation towards the service provider.

One well-known example from Germany is the outage faced by Sky Go viewers in November 2014, where a server problem – possibly caused by a cyberattack – prevented these pay TV customers from watching the Champions League match between Schalke 04 and FC Bayern Munich. Or we could recount the story of the stock exchange going offline in Mumbai, India, in July 2014, when network failures resulted in severe financial losses and claims for compensation – and a significant loss of reputation for the provider HCL.

Extensive root cause analysis conducted on past faults and outages has shown that the typical causes for operational incidents can be categorized into three groups: **people, processes and platforms (3Ps).**

1. Human errors: people

The vast majority of serious faults affecting the operation of mission-critical systems can be traced to human error. Indeed, critical systems can now be given safeguards that make a fault extraordinarily unlikely – as long as a human does not commit a serious error. We can use an example to illustrate this point.

A large corporation suffered a severe service disruption affecting all of its customers in Austria, after the connection between its two data centers in Vienna was disrupted for a period of about eight hours. The cause was found to be a defective cable running between these centers, which had been gnawed to pieces by rodents. While this doesn't sound like a case of human error just yet, the devil is – famously – in the details. There were of course two cables running between the data centers. But the **failover** to the backup connection had not been triggered since the wrong firmware version was installed on the all-important routers. To make matters worse, the failover process had not been tested for some time and a more recent firmware version had not been installed, although it was available and would have rendered the error visible. In addition, the cable ran over the Danube with an increased risk of rodent damage, but the cable was ordered without specifying steel jacketing. Last but not least, the fault occurred at night. If all of the staff in monitoring, technical troubleshooting and management had estimated the criticality of a possible cable failure correctly, the effect on customers could have been minimized and the fault could at least have been fixed before the next working day. In summary then, a cascade of human errors transformed a simple technical fault into a major outage for the company.

2. Process errors: processes

Process descriptions are either unclear or incomplete or old. Since the introduction of **ITIL** (see Chapter 3) and its provision of **standardization and full descriptions of routine processes** in IT, problems with patchy process descriptions have now become something of a rarity. Much more frequent, however, are overly abstract process descriptions that offer little to no guidance on how the tools and system environment should be handled in practice. Other common problems occur when company – or departmental – boundaries are crossed. Overlap or inconsistency in responsibilities can lead to situations where everyone is expecting someone else to step in, resulting in a service disruption. Here, too, we can use another real-world example.

A customer ordered a bandwidth upgrade from a service provider. In technical terms, this meant activating a new, higher-bandwidth line and deactivating the old connection. This process involved the constant crossing-over of department boundaries but, the two technical operations had no process continuity and their timing was not synchronized. As a result, the old, low-bandwidth line was deactivated first and the new connection was only activated later as part of incident processing, with catastrophic results for the customer: multiple total failures relating to **WAN** use, i.e., in the network connections between the company's sites.

3. Technical errors: platforms

As has been mentioned above, technical errors are only rarely the cause of faults in systems with multiple safeguards. However, these should not be underestimated. Modern systems and telecommunications networks are designed with multiple redundancy. In the event of a power supply unit (PSU) defect, for example, a second PSU will take over. The same applies to storage to prevent total drive failure. There are redundant servers, network connections at all layers and other precautions. The probability that both components fail simultaneously is very low indeed. Additional redundancy at a higher level acts to further minimize the risk. This can take the form of a second active and fully specified server in another data center, for example. Despite all of these safety measures, however, technical faults cannot always be avoided, especially in cases where redundancy itself – i.e., component incompatibility – is the source of the problem. Or if a component malfunction occurs only sporadically or in one particular area, generating a "flapping" scenario. Here, service can be disrupted by rapid, uncontrolled activation and deactivation of the failover. For network operations in particular, this kind of malfunction is critical, since all of the downstream services and applications will also fail.

Other frequent sources of error include defective firmware, outdated hardware (networking is a prime candidate here), substandard monitoring that does not offer an end-to-end view, and too much complexity due to the coexistence of heterogeneous technology and version levels. Firmware in active network components is an especially critical factor for example. In large networks, a common problem is that the failover is set up correctly – so that a second component will take over from a failed first component – but certain firmware versions in certain configurations then cause a system-wide failure. In this situation, the backup component either fails to handle the workload, or is so disrupted by the defective component that network traffic effectively ceases. This is an especially insidious problem, since not all possible eventualities can be tested. Furthermore, network outages usually have fatal consequences for the vast majority of services.

2.3.3 Problems at the Interface between the Customer and the Supplier

Good customer communication is crucial for success – particularly in critical situations. Yet this is also precisely where misunderstandings can occur that lead to major service faults even if the supplier is meeting its SLAs. Satisfaction – and its flip side, dissatisfaction – is ultimately a subjective metric.

Here, employees function as intermediaries between the two companies: internal processes must ensure that they receive information quickly and can prepare it for use by the customer. In large enterprises with a number of production units and complex supply chains, however, these processes can be error-prone and time-consuming. Yet the smooth flow of information between operations staff and the **service delivery managers** is critical for ensuring visibility by the customer, and thus for the provider's overall credibility.

Another major source of problems is a lack of access to information from the customer's environment. Service delivery managers are often expected to know the customer's back-end systems – and the roles of the various systems and connections – in detail, and to provide quick and accurate trouble-shooting during a fault situation. They should also be aware of new customer ideas or priorities early enough to translate these into solutions. In a conflict situation, service delivery managers must represent the interests of the contractor whilst, as the customer's representative at the same time, ensure an adequate level of service provision from their own units and escalate delivery grievances internally.

The total sum of these and many other tasks such as **submitting change requests**, offering new services to the customer, or providing consulting services, can place excessive demands on managers, or result in highly complex job roles for which very few staff members are suited. As a result, important activities are either sidelined or poorly executed, and a gap opens up between contractor and client. Over time this can widen into a breakdown of critical proportions in the relationship between the two companies.

ISO, ITIL & Co. – A Baseline and Orientation How-To

How can we guarantee the level of service quality for customers described in the previous chapter and avoid crises? Which existing guidelines are referenced by Zero Outage? And what is the starting point for establishing a zero-defect culture in an IT service company?

A sensible and logical first step to define a mutually agreed and auditable overall framework is to base quality on existing standards. This chapter briefly discusses several such standards and offers suggestions for business practice. Later sections of this book then address these in greater detail, in terms of their role in Zero Outage.

3.1   A Brief Overview of Recognized Guidelines and Standards

3.1.1   International Organization for Standardization (ISO)

ISO standards are published by the International Organization for Standardization (ISO), an independent organization with member bodies in over 160 countries. The basic idea behind the work of ISO is to create internationally recognized standards for safety and compliance that certify companies on the basis of their internal management and safety processes. The core is formed by a series of best practice frameworks, which reference a very wide range of sectors, including IT quality, water utilities, food production and sustainable development. ISO-certified companies obtain documented proof of the safety, reliability and high level of quality achieved by their products and services.

Two possible routes are open to any company wishing to obtain certification. The first – and hardest – route is to read and digest the standard reference works and then transitioning the company as an independent project. But this can be daunting if managers lack experience in the field: even the terminology used can be unclear for the newcomer. This approach is therefore time-consuming and expensive and offers no guarantee of a satisfactory end result.

The second approach is to attend workshops run by accredited inspection bodies or management consultants, and then commission specialists to implement within the company. Whilst more costly initially, it is more results-focused and ultimately more cost-effective, as it avoids any risks of misinterpreting the subject matter. ISO auditors themselves are not allowed to provide consulting services. Whichever route is taken, a company makes a formal application to a certification body when confident of passing a certification audit. These audits assess company compliance with the standard. Pre-assessment audits can also be organized. Here, the company provides a list of parameters not yet audit-ready, which auditors either focus on or ignore for the purposes of this audit. These audits are also limited in terms of depth and scope. A pre-assessment audit should take less than three days, for example, or there could be a suspicion that the auditors are playing a consultant role. The goal is a one-day audit. At the end of the day, the company receives a cost estimate based on company size, number of employees, presence of an internal research and development unit, etc. If the company accepts the quote, the actual audit then follows within a specified period of time.

Once a company is certified, the certification is valid for three years. A re-certification audit is then carried out. After the first audit, two follow-up audits plus recurring surveillance audits are performed to verify that compliance with the standards is being maintained.

Details of the various standards (ISO 9000, ISO 9001, ISO 20000 and ISO 27001) are provided in the Annex.

3.1.2   IT Infrastructure Library (ITIL)

ITIL is a widely used best practice framework whose level of acceptance has made it the de facto standard in its field. ITIL is a five-book compendium defining the ideal approach towards managing IT operations. Initially developed in the 1980s, ITIL was first published by the UK Office of Government Commerce (OGC), a British government agency.
The individual volumes are:
- service strategy (SS)
- service design (SD)
- service transition (ST)
- service operation (SO)
- continual service improvement (CSI)

Each book illuminates an area of service management in an IT context.

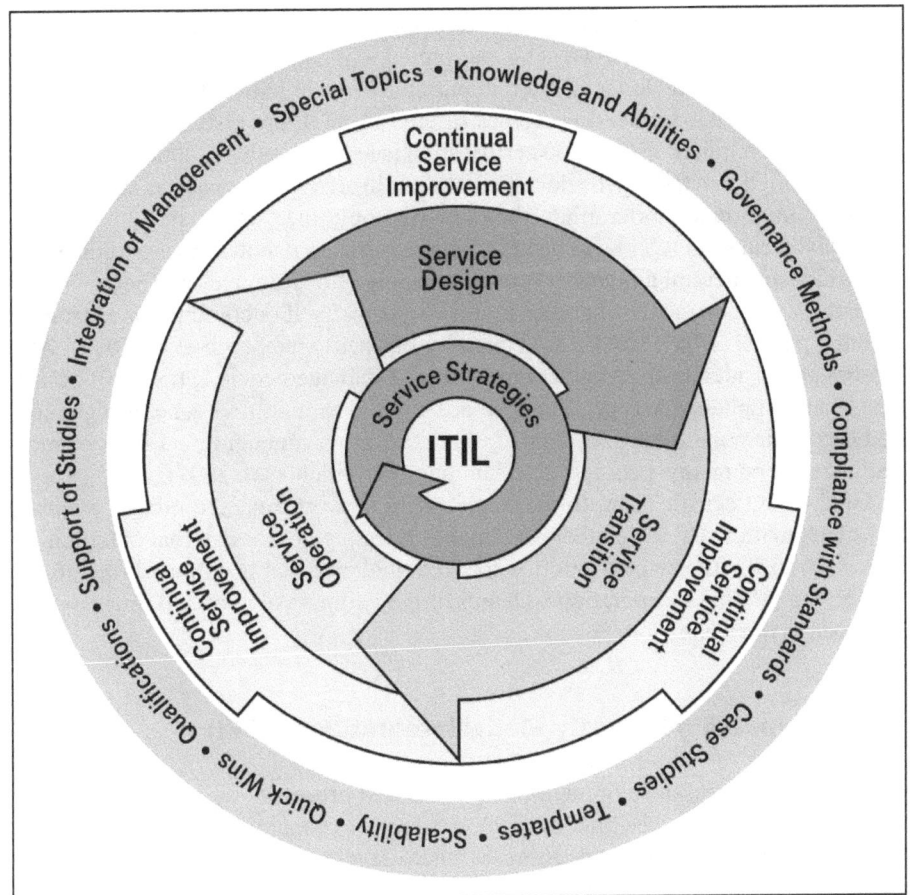

Fig. 3.1   ITIL Life Cycle
© Itil V3 Volume 1 (slightly adapted figure)

Service strategy addresses the necessity for IT service providers to prepare for the eventuality that they can be undercut by other providers offering cheaper and/or more efficient services. Providers therefore need to continuously improve their competitiveness and scrutinize their own performance as an ongoing process. service design addresses the product portfolio of individual clients while service transition then looks at the task of implementing business requirements as IT services. Service operation focuses on avoiding disruptions to process flow after service roll-out, while the continual service improvement volume offers an overarching view of the continuous optimization of products, service provision, quality and processes.

As Fig. 3.1 shows, the separate volumes collectively form a coherent, self-contained framework that offers an excellent foundation for companies looking into redesigning their own IT to be more efficient. At this point, it should be mentioned that ITIL is intended to act as a valuable reference and does not claim to be the definitive guide to service management. Nor should ITIL be viewed too literally. The contents of its books must always be interpreted within the context of the company. Implementing it to the letter is often impossible or even harmful for the enterprise concerned. The standardized processes offered by ITIL are also not sufficient for obtaining the required ISO 20000 certification (see Schiefer et al. 2008, and the Annex). Although purchasing the documentation is itself a minor expense, the volumes have one important shortcoming: they describe only the "what", not the "how" – primarily because ITIL aims to be of use to many different enterprises and therefore offers industry-neutral guidance. Transitioning the IT departments and acquiring the know-how to do so must be handled by the company itself or commissioned from an external partner, which can be an expensive business (see Stych et al. 2008). Also worthy of consideration is "that company and IT management […] [need to] ensure, first, that compliance and performance are tailored to modified general regulatory and economic conditions and, second, that continuous and continual improvement optimize the company's competitive standing" (Fröhlich et al. 2007).

Unlike ISO certification, for example, there is no recognized quality standard associated with ITIL, which therefore makes it less useful for external communication. Instead, ITIL provides practice-focused guidance that is extremely helpful in optimizing internal IT operations. Details of the various volumes and processes are provided in the Annex to this book.

3.1.3 Capability Maturity Model Integration (CMMI)

CMMI supports the development of products and processes by assigning these to various "capability levels" that describe the degrees of maturity. This enables identification of those products and processes where development is still required. CMMI is a universal standard that can be applied to any company operating according to ISO/IEC 15504, and enables comparisons to be made both between products and processes within a company and between one or more other companies.

CMMI uses a cascading system of ascending maturity to define the following five capability levels:

- **1 Initial:** Processes are neither coordinated nor defined. The company has no overall business plan and success depends on individual factors such as employees happening to have certain skills or abilities.
- **2 Managed:** Processes are planned, controlled and assigned resources.
- **3 Defined:** Processes are standardized; standards are documented in writing before rollout, and compliance with these standards is monitored. Process design is in itself a continuous process.
- **4 Quantitatively managed:** Statistical methods are used to control processes. Quality, performance, efficiency, and effectiveness is managed.
- **5 Optimizing:** The statistical data obtained are fed into deviation analysis used to drive long-term process optimization.

A continuous data display format is also used, although only the cascading maturity system is used for certification and definition of a CMMI level. Assigning and confirming a level for an organization involves a highly detailed assessment stage, in which all aspects of each process area are investigated for a corresponding level of maturity. This investigation requires fully documented results to be presented from the affected organizational units and the relevant projects. These are collected in interviews and from document analysis in a process that typically takes several weeks. Any critical deviation is an obstacle to certification. The official assignment of the CMMI level is made solely by persons authorized by the Software Engineering Institute (SEI). The SEI is the R&D unit at Carnegie Mellon University in Pittsburgh (Pennsylvania, USA) originally responsible for developing CMMI. Development work is now handled by the new CMMI Institute.

3.1.4   Lean Management and Six Sigma

As the name suggests, **Lean Management** is a management style characterized by streamlined, optimized processes and high efficiency. Various production methods, such as Henry Ford's continuous flow, Ohno's Kanban system at Toyota or the familiar Deming cycle (Plan-Do-Check-Act) were investigated by researchers at the Massachusetts Institute of Technology (MIT). The results were then used to derive the theory of Lean Management (see Gorecki et al. 2014).

Lean Management aims to provide a solution for the following: "The wastage of resources due to activities that do not create value, the simplification of processes and workflows, and [...] continuous [...] improvement" (see Gorecki et al. 2014).

The successes – both potential and actual – attributable to Lean Management include:

- "Reducing inventories (semi-finished goods, finished goods) by over 50%,
- adjusting lead times to customer requirements,
- substantial improvements to product/service quality,
- reductions in rejects and defects in production,
- increasing customer and employee satisfaction,
- significant improvement to financial key figures, such as return on investment" (Gorecki et al. 2014).

To resolve the triple challenge in production and service management of balancing the competing factors of product quality, lead time and cost, the "elimination of waste" should be looked at in greater detail (see Gorecki et al. 2014). "From the customer's perspective, wastage is anything that makes no contribution to value. Wastefulness ('muda' in Japanese) is a key concept in Lean Management" (see Gorecki et al. 2014).

Lean Management follows five principles to achieve its goals:
- "Precise description of the value of the product or service. The value of a product or a service is defined solely from the customer's perspective. For

companies that want to realize Lean Management, the customers' requirements and their 'appreciation' of products or product properties are therefore given the highest priority.

- Identification of the value stream of the product or service. This refers not only to the internal supply chain, i.e., the value creation process flow within the company, but also to the overall network of companies (suppliers, suppliers of the suppliers, etc.) that is responsible for the manufacturing of an end product (external supply chain). All activities (whether value-creating or not) form part of the value stream and are thus a part of Lean Management.
- Flow of value without interruption. This lean principle calls for a value creation process that is not interrupted by the storage of intermediate and end products or idle time in the production process. This principle is the most difficult to apply, since the batch-driven processing of interim steps (batch manufacturing) seems to be a kind of unwritten law in human labor that yields only reluctantly to the value flow principle.
- Pull of the value by the customer. In accordance with this principle, the value stream is not set in motion by the planning process at the manufacturing company but by requirements or demand at the (end) customer. By applying the principle, production occurs only when products or services are required.
- Striving for perfection. Just as learning never ceases in our knowledge-based society, so too is Lean Management a never-ending task. The implementation of the flow and pull principles can always be improved, and even companies such as Toyota, which started putting Lean Management into practice over 50 years ago, continue to strive towards perfection" (Gorecki et al. 2014).

In the ICT industry, the primary goal when applying this methodology is to identify activities and process steps that waste time and should therefore be optimized accordingly. As one example, the periods of time in which employees typically wait for a backup to finish as part of a software upgrade could be usefully spent in planning a new task.

Six Sigma is a methodology that draws on the application of statistical and analytical methods to improve processes and process output. These are precisely the aspects in which it differs from other process optimization models: it is mathematically based and represents every process as a mathematical function.

Six Sigma consists of five phases, termed the "DMAIC cycle":

- Define: What is the problem?
- Measure: How can it be measured?
- Analyze: What are the causes?
- Improve: How can it be eliminated?
- Control: How can I avoid this problem in the future?

Application of this methodology always involves the assumption that there is a problem to fix or a situation that is to be improved. Accordingly, the first step is to analyze the genesis and type of problems that occur to then eliminate them in a targeted and controlled manner (see SixSigma 2015).

While Six Sigma is conventionally applied to the optimization of production processes in manufacturing, its methodology of identifying the right key figures for control and for measurement of the same is also very applicable to the ICT industry. As one example, measuring the time taken for each process step is an indispensable part of being able to control productivity. In this way, starting values can be identified for process chains and then be improved in a targeted manner by comparisons between separate organizations, with customer requirements or the latest benchmark. The underlying premise here is always to achieve no less – but also no more – than 100 percent of the stated objective. Further details, covering rollout in the enterprise, opportunities for skills training, and the Lean Six Sigma variant, can be found in the Annex.

3.1.5   Project Standards

Following the presentation of the various general enterprise standards and frameworks above, the next sections now turn to look at standards used in project management.

A project is created at the moment in which a company decides to take some sort of action: to implement an idea, to develop something new, or to make an improvement. This project is assigned to employees and the whole undertaking is given an entrepreneurial structure. In this way, the same kind of structures are generated within the project as are found in company hierarchies: with one or more managers at the top management level and staff members executing the project in their various roles. Depending on the project size, additional management functions may be added underneath the top management layer – such as for heads of department or team leaders. The project standards below are intended to show how projects can be organized to be target-oriented, controlled and coordinated. They are therefore recommended as a guidance.

3.1.5.1   PRINCE2

PRINCE2 is a tool for ensuring that projects are managed in a coordinated and efficient manner. Written out, the acronym expands to "**PR**ojects **IN** a **C**ontrolled **E**nvironment". This project standard was established by the Office of Government Commerce (OGC), a UK government agency, and continues to be maintained and further developed by this body. It has been deployed in over 56 countries and is a familiar concept in many companies worldwide. The standard has existed since 1989. In 2005/2006, the revised version PRINCE2 was published, and this was again revised – and simplified – in 2009. Since the underlying structure remained the same, however, the name was not changed to PRINCE3 (see Beims et al. 2015).

PRINCE2 is used to handle parts of project preparation, the project start, completion, and a coordinated end to the project (see Ebel 2011). Further details on the structure of the standard are provided in the Annex to this book.

3.1.5.2 PMI

The PMI (Project Management Institute) has been responsible for the world's leading standards in project management. "PMI is the largest international industry association for project management [...] [with] over 640,000 project management professionals (PMP) worldwide (approx. 12,000 in Germany)" (Kayenta 2015). "PMI global standards provide guidelines, rules and characteristics for project, program and portfolio management. These standards are widely accepted and, when consistently applied, they help you, your global peers and your organization to achieve professional excellence" (PMI 2014).

PMI standards have the advantage of being industry- and market-neutral, and can also be used independently from the user's educational background. Anyone can gain certification. The standards are clearly formulated to be understandable for any user – economists, IT professionals, and career changers alike. Annual surveys among project managers on day-to-day work, changing requirements and practical know-how ensure that PMI is a self-learning and continuously improving product. Here too, the focus is squarely on best practice – in the sense of a guideline by project managers for project managers (see PMI 2014).

3.2 Practical Applicability of the Standards

Working according to defined standards is a key criterion for Zero Outage. To determine which of the frameworks described is most applicable, the first step should be to assess the expected benefit for the company along the following lines:

- What goals are pursued by the environment? What do suppliers and customers expect from the company in terms of quality? How is the competition positioned?
- Which (technical and business) goals are pursued by the company? How does the company view quality as contributing to business performance?

3.2.1 Considering the Environment

Let us look at the first question: the environmental analysis. Our environment will generally affect the actions we take, since we must either respond to the needs and activities of our environment or – ideally – should be prepared to handle them. Caution is advisable in internal quality management when suppliers or customers become involved in processes: to ensure that integration proceeds as smoothly as possible, the analysis should already have considered the issue of compatibility.

What relationships and factors are in play between suppliers, customers and the current competitive situation?

Supplier Relationships

What is the structure and scope of the portfolio offered to us by our suppliers? What is the priority – low price or service quality? Obtaining an accurate evaluation of suppliers is a key factor for designing the company's internal quality processes, the selection of suitable methods and the integration of suppliers into processes and internal quality assurance. ITIL offers a number of options for this task.

Customer Relationships

The three crucial questions are: what do our customers require of us, what do they get from us in real terms, and what are we prepared to give them?

First, we need to understand our customers to compare their expectations with what we are offering. Six Sigma vividly describes this as the "voice of the customer". And this is the entire point: our opinion of what the customer wants is not the decisive factor. Instead, we need to listen and give clear feedback about what we can provide. What we offer must meet the customer's demands. It doesn't necessarily have to improve on them, but must never fall beneath this standard. Why? To put it simply, customers are always mindful of the price that they are prepared to pay – and for quality in particular. If we exceed requirements, we may be too expensive. If we set our bar for quality and effort too low, our prices will be attractive generally, but we will fail to meet the customer's standards.

The Six Sigma methodology and toolbox is the reference work of choice for analyzing customer relationships, and for choosing an appropriate orientation and control system for quality production.

Competition

When analyzing the competition, studying and identifying market trends is essential for ensuring that our next steps can be planned accurately. Offers already present on the market must be compared with internal performance and the differences be established clearly. Only in this manner can we find out whether our own products and services match market requirements – and can perhaps even be identified as key differentiators. CMMI can be used to assess the maturity of existing products at this stage to highlight possible further actions that can be taken. Awareness of the conditions governing the competition is also important, e.g., the required ISO standards, competitor certification, and what counts as industry best practice. Consolidating this information enables the definition of the company's competitive position and a target position, and the creation of an action plan.

3.2.2   Identifying Internal Motivators

Now, to the second question: what are the (technical and business) goals we pursue? How do we view quality as contributing to the company as a whole? Achieving compliance with a standard for the sake of compliance is not a path to quality. If we want change, we need to choose and formulate our goals wisely and tailor them to the current market situation. Indeed, we position ourselves with our goals: we can choose to meet or even exceed the needs of our customers for example, and we can operate at the conventional level of technical expertise or set our sights on staying ahead of the market.

While this course of action determines the general approach to follow, it must subsequently be reviewed from a business perspective as well. This is because quality initially involves costs. But Zero Outage in particular attempts to improve efficiency by utilizing standards, thus leading to cost reductions. Expenses and expected results must therefore be weighed up from an entrepreneurial perspective.

3.2.3 Choosing the Right Strategy

After proper consideration of the general conditions and the identification of internal motivators, the next step is to draw up a list of priorities and implement the project. At this stage, the results of individual analyses must be seen as pieces of a puzzle that must now be put together to form an overall picture.

It is not necessary to apply every single method religiously. Instead, the correct pieces must be identified for our own company and be placed into position in the final puzzle. The aim should be to bring together the standards and tools from the various frameworks that best fit together – even in a modified form if necessary. Absolute 100 percent adherence to a standard is sensible only if this is expressly called for. If not, then one should pick out the parts that are most useful and integrate these according to our own needs.

Focus on Quality – Trends and the Realities of Business

4

Now that we have looked at the various standards and frameworks, we can turn to the question of where the future is taking us in terms of quality. How do companies position themselves strategically for the future? Various studies, analysts' assessments and publications from well-known companies and stakeholders with a practical interest in these matters have pointed out significant trends in the role and significance of quality. But where in the day-to-day operation and organizational structure of the ICT provider should the increased importance of quality be reflected?

4.1   Studies Show: Quality Is the Key

A number of insightful studies have been published on this subject.

For its IT Sourcing Study in 2015, PricewaterhouseCoopers (PwC) interviewed over 50 IT outsourcing service providers with a significant market share in the DACH region (Germany, Austria, Switzerland) and the Netherlands. More than two out of three respondents said that the quality of services would be a decisive criterion in their choice of suppliers in future tenders (see PwC 2015). This concurs with an earlier IT sourcing study which asked customers which characteristics they considered particularly important in an IT service provider. Again, quality was mentioned as the number one criterion – even before price (see PwC 2012).

In mid-2015, the Information Services Group (ISG) carried out a study for T-Systems that looked at what motivated decision-makers when selecting suppliers. "For 97 percent of companies, IT quality is now either 'highly' or even 'very highly' critical to the success of individual business processes" (ISG 2015). But is quality really a purchasing criterion? "98 percent of respondents said that IT quality was 'very often' or 'always' a factor in the company's decision-making process. General performance (including the stability of processes and reliability of services) and previous references were other important factors" (ISG 2015).

In a 2015 study about IT trends, Capgemini confirmed the continued growth in the use of internal corporate clouds and in the demand for near real-time scalability of ordered services. The subject of security is becoming an increasingly important part of strategies for safeguarding business processes and protecting against industrial espionage. This is not least because system networking, as used in digital supply chains and supplier management, for example, is on the increase (see Capgemini 2015). As security cannot be achieved without very high quality standards, this study's findings also confirm the importance of quality.

All of these studies pointed out the need to make quality a priority.

4.2 Running a Business: Seeing the Big Picture .

The logical consequence of the tremendous importance that customers attach to quality when selecting a supplier is that quality should be a central theme of the business and not just a matter for project management or new business development. Seeing the big picture is extremely important to businesses who want to be market leaders. But how can an ICT provider ensure that quality is an integral part of day-to-day operations?

4.2.1 Focusing on Solutions That Work Long-Term

Disruptions can have unacceptable consequences for your customers' daily operations. They mean extra work, delays, and maybe even outages. That's why it is so important to learn from previous disruptions – to prevent them from occurring again. And because architectures are becoming increasingly complex and efficiency improvements are constantly being required, companies should have a problem management system in place. This will then not only look for a permanent solution to the disruption, but will also document workarounds that can be used in the interim. What is more, it will reduce recovery times dramatically should the problem ever reoccur. The Materna Monitor has reported an increasing focus on problem management and emphasizes the importance of deploying resources wisely and sticking to the rules (see Görgen 2015).

4.2.2 Thinking End-to-End

Thinking end-to-end about something means going back one step at a time to gain a broader perspective. The following example shows how this works.

When an ICT service provider manages a system for a customer and wants to deliver maximum quality, it must also understand what the consequences of a system failure would be in a broader context. To do this, the service provider must understand the big picture. This means, for example, being aware of other systems (currently not affected) being managed for the same customer. Some of them will be interacting with the first system, and some of these linked systems might even be administered by the customer himself. It should be clear, therefore, that the systems managed by the provider are not the only ones involved in the day-to-day running of operations, because there are also other systems that are operated by the customer. Or the line that links the systems was sold and installed by the provider as part of the deal, but it may be operated by a supplier. Therefore, the provider should define all processes in cooperation with his supplier.

This example is typical of a wide range of systems and services. Therefore, a "landscape" should be created containing all existing components from the customer right through to suppliers. Although this is still in its infancy, there is a noticeable trend towards the creation of landscapes. Since sourcing strategies are becoming increasingly complex, responsibilities are shared by different partners. And if the aim is to achieve the highest standards of quality, these partners need to collaborate effectively. Ultimately, the customer expects a top-quality product at all times because that is what he is paying for.

4.3   In the Organization: Putting Quality on the Right Footing

Now that we have looked at the studies and the changes in the way businesses are run, we can take a closer look at how quality is achieved, and how the process should be organized and defined.

4.3.1   Prioritizing Quality – Giving It More Importance in Operations

More and more companies are using and applying the frameworks and standards discussed in this book. There is evidence that quality management is becoming an increasingly important part of operating activities in IT and telecommunication companies. But simply offering advice on the subject is not enough. Quality needs to become a major concern for the organization. And it needs to be championed at an appropriate level within the organization's hierarchy, with senior management playing a coordinating role. Typical of such an approach would be the active management of disruptions. To deal with service outages that affect more than one user, a cross-departmental approach is advisable – one that is on the same hierarchical level as the production unit in order to execute and monitor the processes.

4.3.2   Broadening the Definition – Establishing Quality Internally

Customers place particular demands on the services provided by their suppliers. Quality is therefore defined by what the customer expects. The increasing market penetration of cloud services, digital processes, shared platforms and more has had the effect of drawing greater attention to the issue of security.

Security contributes substantially to the customers' perception of their providers' quality. A solid, proven security strategy as an expression of the supplier's quality focus has an equally positive effect on the customer's success as a company. Some vendors have already acknowledged these customer concerns and feel duty bound to meet them. It is not unusual for vendors to adjust their own organizational structures to meet these requirements by, for example, setting up their own cross-organization security units.

4.3.3   Thinking ahead – Creating Alliances

The move towards end-to-end thinking is also reflected in the way the quality function is organized. Recently, companies within the same industries have joined forces to provide their customers with even better quality and greater efficiency. Here, service providers and partners communicate with the customer and with their suppliers. The different areas of expertise and strengths that each member brings can deliver a better outcome or a more comprehensive service for the customer. These positive results serve to increase demand, so that all of the parties involved benefit from the alliance.

A good example of this kind of alliance is ngena (Next Generation Enterprise Network Alliance – see http://www.ngena.net/), a global alliance of international network providers, including founding member Deutsche Telekom. Starting in the first half of 2017, this independent company intends to provide services to international business customers – with all the alliance members providing network access in their own markets. ngena will link the partner networks to a global network, making this available as a platform to members. This "sharing economy" business model is unique thus far in the enterprise network business. ngena's aim is to provide international network services offering high security, flexibility and quality to corporate groups and SMEs. The virtualization of network functions and end-to-end automation will make the services particularly efficient. ngena will also enable companies to connect quickly and easily to remote locations in the future.

Building a Solid Foundation: The Four Cornerstones of Quality Management

To keep up with – or even create – trends in quality management, every company should keep an eye on its own internal situation and targets. Not least because all systems and companies need to focus on a long-term mission, from which they can derive a strategy, and its day-to-day and tactical implementation. At regular intervals, companies should be asking themselves "What's our blueprint – and what do we want to build?" and "Have we laid the groundwork properly?" ICT service providers should also routinely question the status quo. Particularly since the provider's portfolio (infrastructure hosting, software solutions, etc.) obviously must turn a profit, the question arises as to how this profit can be achieved and the principles the company should adopt to support this goal. Should the ICT service provider emphasize portfolio quality and also offer custom solutions? Or is large-scale standardization and cost-effective services a better approach? Is the sweet spot somewhere in-between? And how is the company organized to support all of this?

The International Organization for Standardization (ISO) has published a guideline that helps companies establish a solid foundation for their quality management strategies, in the form of a document outlining quality management principles (QMPs). This useful publication is available at no charge (see International Organization for Standardization 2015).

The document introduces the eight principles on which the ISO 9000 **quality management system standards** are based. The principles are described in greater detail in ISO 9000:2005, "Quality Management systems – Fundamentals and vocabulary" and ISO 9004:2009, "Managing for the sustained success of an organization – A Quality Management approach".

The principles given in ISO 9004:2009 are:
- Principle 1: Customer focus
- Principle 2: Leadership
- Principle 3: Involvement of people
- Principle 4: Process approach
- Principle 5: System approach to management
- Principle 6: Continual improvement

- Principle 7: Factual approach to decision making
- Principle 8: Mutually beneficial supplier relationships

In this chapter, we will use these principles as models for defining four cornerstones we believe are essential to quality, and which our experience has shown us offer the greatest leverage for improving IT quality management. We will discuss successful partner and supplier relationships in the context of Zero Outage in Chapter 14.

5.1 The Maxim of Customer Focus

As we never tire of repeating, quality is never an end in itself. This naturally also applies to quality management in IT, where customer satisfaction is the only metric – perhaps reminding us of the old saying "You not only have to be good, you have to be seen to be good." Seems unfair? Perhaps. But it illustrates the importance of the customer's perception.

IT companies must understand the current and future needs of their customers in order to satisfy their requirements and intentions. A long-term partnership of equals between customers and ICT service providers typically takes years to mature and can only flourish if customers feel they are being listened to and are understood when talking about their specific business requirements.

These individual requirements will naturally vary from customer to customer. Take an international manufacturer of elevators, for example: the company sells its products worldwide and also has maintenance agreements in place. This also applies even in China, for example, where the elevators have been installed in various provinces many miles apart but need to be inspected every two weeks to comply with legal requirements. The elevator company therefore has a clear-cut problem: it needs to meet its maintenance obligations with a digital solution to conserve resources and save the seven-figure sum it would otherwise spend on on-site visits. The ICT provider needs to focus on solving a tangible business issue for the elevator manufacturer. Only then should additional services be offered – perhaps from the provider's cutting-edge solutions portfolio. Overall understanding of the customer is essential when creating the solution to a specific problem.

Business success can only be achieved if the provider's internal organization and quality management are systematically oriented on customer requirements – together with rising market share, improved customer loyalty and follow-on orders.

This principle also implies two key internal activities:

- Routine measurement of customer satisfaction and taking action on the results
- Systematic customer relationship management

Sales staff should not be the only ones familiar with the customer's requirements. These should be systematically communicated throughout the organization as well – including product and portfolio management. The provider should also ensure that its goals are closely aligned with customer requirements in both breadth and depth. Last but not least, stakeholders should also be considered. Aligning customer interests with the interests and resources of other parties – such as employees – is especially important.

5.2 Involvement of People and Leading by Example

This brings us to the most important asset for any organization: its people. And by "people" we mean employees in all parts of the organization, not just management. To organize an IT provider along Zero Outage lines, employees are required who can develop a deep understanding of this quality approach. Quality needs to become part of the corporate DNA.

Employees must therefore become actively involved, for example in the strategic development of the quality unit or the introduction of new process steps. The purpose and intention of new methods should be explained, not dictated: to create meaning with knowledge.

We all recognize this from our own experience of corporate life and simple human nature. If we are involved in decision-making, we are more likely to support it – even if the outcome is initially difficult or uncomfortable. One example is the introduction of new change processes that require greater documentation or coordination effort. "Keep discussions open but decisions focused." Satisfying this entrepreneurial principle usually requires good communication among staff.

Employees who see how their work contributes to the success of the business in an overall context will also be

- more creative and innovative in supporting the organization's goals;
- fully aware of the value of their own work and the effects that it has – whether positive or negative – on the organization and its quality goals;
- more likely to participate in the continual improvement process and share their knowledge;
- eager to expand their competencies and knowledge; and
- anxious to encourage and promote an open-minded approach to problems and weaknesses.

In IT quality management, continual improvement and its associated constant change are part and parcel of day-to-day business. Effort should be made to focus employee motivation in both hearts and minds.

Management culture also has an important role to play in the achievement of quality goals. The management team is the most important advocate for strategy and sets an example for staff to follow. Later we will be looking at how far this management-as-role-model concept can be developed – for example in scenarios such as major incidents where management attention is needed to speed up problem resolution. For now, suffice it to say that management's attitude towards quality and quality requirements must be authentically reflected in their own actions and behavior. In addition, employees need to receive consistent statements about the nature and importance of corporate goals, wherever they work in the organization and regardless of the management grade providing this information.

Frequently, companies deliberate about how best to provide their employees with the support and training they need for specific processes or tasks. However, these efforts often ignore the management staff themselves, who could already be facing additional challenges, such as team leaders in IT operations, or who are trying to build support for a proposal. So, one important step is to ensure that the needs of this group are adequately understood, and to take these into account when preparing information packs and training courses.

5.3   Continual Improvement

The continual improvement principle is our constant companion on the path towards achieving the vision of Zero Outage. Importantly, improvements must be permanent, and can and should be made throughout the entire company: in quality, in operations, in project work, in customer interaction, etc.

The question therefore is how to prepare your entire organization for it and how to turn other divisions into catalysts for improvement as well. From experience, we recommend that the quality team should share its vision of improvements and its key quality initiatives with other units in the company and ensure that all of the relevant divisions back this vision. In larger groups or companies, you will be unable to make much of an impact without active support from specific stakeholders and thought leaders. Furthermore, it is more likely that project teams themselves will be the source of sound improvement recommendations enabling the completion of projects on schedule and at the desired level of quality.

In many cases, ensuring that changes address the right issues and are implemented appropriately is more important than being the leading champion of all key initiatives. For example, a new automation project that enables a highly efficient, automated workflow for server installations will automatically reduce the number of installation faults due to human error. This positive side effect also pays dividends in terms of quality and should therefore be recorded in a suitable "Human Error" **key performance indicator (KPI)**. Such KPIs provide a system of metrics that help you measure the current status quo and any quality enhancements achieved through improvement measures. Not all projects can be executed following a decentralized approach. Some – like the one mentioned above – are better suited to being run by a joint steering committee (e.g., in IT operations) with an appropriate reporting line.

Another important step here is to ensure that the relevant initiatives for improvement and the result itself are fed back into a centralized quality unit. This core unit can then support the Zero Outage vision in terms of the three factors of people, processes and platforms (3P – see Chapter 2).

As a last point on continual improvement and the need to consider the bigger picture: train your staff or your core team not just in terms of IT quality management but also as regards methods and tools of continual improvement. And reward the improvement of products and processes rather than maintaining the status quo (via agreed targets, for example).

5.4 Factual Approach to Decision Making

Effective and goal-oriented decision making is not the result of instinct but is based on information and data instead.

Decisions based on objective facts

- mean real data can be used to demonstrate that past decisions were appropriate and correct; and
- strengthen one's own position when lobbying for a specific decision or attempting to convince key stakeholders, while ensuring one's own actions can be analyzed later.

For example, let's consider the rollout of a quality program or set of international process standards, which will take between one and x years depending on the size of the company and will tie up a large number of resources that are in great demand. The clearer the **facts, figures and data** on which this decision (and any necessary "course corrections") has been based, the more likely it is that the initiative will receive broad-based support and achieve the desired outcome. And three years down the road (for example), how satisfactory the following statement will then be: "We have reduced critical faults by 30 percent, while also – as a direct result – increasing customer satisfaction by 40 percent!"

Remembering that the most important factors for quality are discipline, reliability and attention to detail, these can also be applied to the underlying facts and figures that should inform our decision making:

- Information should be sufficiently available, current, correct and reliable.
- Ideally, data collection should be automated and involve little manual intervention (and human error).
- Data should also be available to all of its consumers at all times (e.g., via a portal with export functionality).

Those with extensive experience in quality work will know that, while numerous things can be measured, there are also plenty that cannot be reduced to useful KPIs or be used as a clear indicator of important relationships – such as customer satisfaction versus SLA fulfillment. Initially, KPIs should be kept to the absolute minimum and specific to those most meaningful for accomplishing the Zero Outage mission.

Quality in the Organization: From Individual Functions to a Zero Outage Organization

6

An effective **organizational structure** provides the foundations for the successful implementation of a corporate culture that ensures quality and customer satisfaction. The critical success factors are explained below. We present several typical organizational structures with their individual advantages and disadvantages and then describe the Zero Outage organization that has proved to work.

An organizational structure is only one of the criteria needed to fulfill corporate objectives; it is not the be-all and end-all. While designed to help organizations achieve their primary and secondary objectives as efficiently as possible, it can never replace key personnel or make up for poor implementation. Similarly, individuals in the organization can be successful only when the structure supports them.

6.1 Objectives of the Quality Function in the Organization

You might have come across "Wayne", a full-bearded colleague in Birkenstock sandals who trundles into the office around 10 o'clock, calmly puts on a brew, watches the tea draw for a full seven minutes and doesn't even take off his coat during that time. He then switches on his PC and waits calmly and leisurely until all programs have booted up. If we ask Wayne for help, he will answer, "If you had listened to me last year and implemented the new XY framework earlier, you wouldn't be in this situation now." He then goes off on a tangent, embarking on a discourse on the introduction of hard disks – and as he drones on, everyone has already mentally switched off and is back working on the urgent problem that the customer is now threatening to escalate to the executive board.

In actual fact, Wayne has a great deal of expertise. He works in the methods and standards department, a cross-divisional function. Or in architecture. Or in quality. We know that these areas need people who optimize things. On the other hand, we also know that Wayne will never really improve things. This is because his practical knowledge of the problems is limited. What is more, he does not create solutions.

And, most importantly, no one listens to Wayne. Even if they do, no one does what Wayne proposes. This is because there are things that are extremely important and urgent that need to be done first. Every single day. It is a vicious circle that must not be allowed to start in the first place.

To effectively support customer satisfaction and quality in an organizational structure, two objectives must be achieved: firstly, **permanent acceptance of the quality function** in the organization, and, secondly, an organizational structure that provides the necessary influence for this function. The human resources function is the best example of this: if an important quality project is due to be implemented with the customer, the priority of other tasks may need to be changed in order to free up the required resources.

6.2 Positioning Quality Correctly

Cross-divisional functions in IT such as those in the architecture, process management, quality or procurement divisions often have the reputation of being impractical and sitting in an ivory tower. The staff and the management team in these areas don't directly experience pressure from the customer. As a result, direct contact with daily operations is lost, and over time individuals who used to be recognized as highly qualified experts become perceived by operations managers as "academics" who are far removed from reality – much of what they say is correct but cannot be implemented in everyday operations.

The objective of establishing a quality function in the organization must therefore be to act in a practical manner all times and ensure that the improvement initiatives and the operating processes and quality assurance procedures are of use when problems arise. This is the only way to ensure that staff from the quality organization will find high acceptance in other units, people will not work against each other, and no parallel organizations will be established.

Where there are plans to implement the Zero Outage method effectively in an enterprise, an organizational structure is needed that enables the quality manager not only to develop but also to implement quality assurance concepts and initiatives. For this, it is essential to ensure **hierarchical transparency**.

In practice, quality enhancements always involve change. What this will mean for the company and its staff is naturally impossible to predict. As the philosopher Georg Christoph Lichtenberg eloquently put it: "I don't know if things will get better when they change. But things must change if they are to get better."

Every change leads to resistance in an organization, and every change initiative creates additional work for a lean team that in many cases has already focused on improving efficiency on several occasions. Consequently, many senior managers from the line organization initially see each new requirement and each quality improvement project as at odds with the overall requirements for further cost-cutting and efficiency enhancement. A quality manager without an effective escalation instrument will therefore fail in any organization.

The following section shows how quality tends to be positioned in IT organizations.

6.2.1 Typical Positioning of Quality in IT Organizations

As a Staff Function Headed by the IT Director/CIO

In this organizational structure, the quality department comprises of a handful of staff members who mainly set regulations for frameworks, tools and methods, without considering how the content affects the operational and project teams. The biggest worry here is that sooner or later this department will get the reputation of sitting in the above-mentioned ivory tower – whether justified or not. This is because the employees of this staff function are not involved in operations. This fact alone makes them "suspect" in the organization. Ultimately, the acceptance and consequently the success of this unit will suffer.

As Part of the Regular Line Organization

Every organizational unit has its own quality function that imposes specific regulations for the unit in question (e.g., application development) and performs checks. When inconsistencies occur, the issue is escalated to the line manager (e.g., the head of application development). Regulations on standards, tools, processes and frameworks to be used are line-specific but a good fit with the requirements of the relevant unit. Improvement initiatives are generally driven by new issues and are implemented by the organizational units. One of the disadvantages of this form of organization in large companies is that standardization suffers as different line-specific examples of the core processes and procedures arise over time. What is also missing in such a setup is an unbiased view on quality that is independent of the respective line manager.

As a Cross-Divisional Function

Here, quality is defined as a cross-divisional function for line organizational units, such as server management, network management, desktop management and application development or application management. The quality function comprises process managers, project managers, and specialists in security, compliance and frameworks. Staff members continuously measure quality using fixed KPIs that apply across the entire organization, launch improvement projects based on a fixed methodology (such as Six Sigma) and turnkey regulations into standards for documentation, security, tools, test methods and process models. The drawbacks are similar to those of the staff function.

6.2.2 The Zero Outage Organization

So, what characterizes an effective setup that ensures that practical regulations are created and improvements are implemented?

Firstly, the structure of a Zero Outage organization must be centralized enough to ensure standardization at a technical, procedural and behavioral level. Secondly, it must have a detailed understanding of the problems of the operating units to draw the right conclusions. This is the basis for implementing improvement initiatives. That this can give rise to conflicting objectives is a logical result.

The Zero Outage organization is a **combination of the organizational structures outlined above**. It combines the advantages of the relevant variants and evens out the disadvantages. The following factors are key for this:

a. **A corporate team** that reports directly to corporate management and that comprises an operational area that interacts directly with a conceptual and strategic area. The operational part of the team leads the way in the solution of serious incidents including follow-up in problem management, supports risky, complex changes and de-escalates complex projects that run into difficulties. It is important to define from which risk category or "customer impact" the operational units must hand over management of a change, incident or project to the corporate quality team. The conceptual arm of this corporate team (ideally the same people) then analyzes the operational problems to develop the resulting improvement initiatives in the same way that measures designed to prevent an incident from recurring are defined in problem management. This ensures that the quality initiatives come from practice and are suitable to eliminate the incidents that occurred. The corporate quality division owns the processes that are critical for quality in accordance with ITIL and project methodology. By conducting internal audits, it ensures that the same processes and terminology are used throughout the company.

b. **Local quality teams in the operational production units:** They advance technology-specific standardization in their respective areas. In addition, they ensure that technological sources of error are systematically evaluated and that any resulting corporate projects aimed at making improvements at a technological level are implemented. The local teams handle the regular incidents, changes, problems and projects, normally as part of or in conjunction with the operating units.

c. **Local teams in the service units:** They ensure that customer-based regulations and improvements mandated by the corporate quality division are also regularly implemented in service management. Likewise, they make sure that the information about the customer's system landscape is up to date and documented. In regular meetings with the customer (service review boards), they collect information about any problems and dissatisfaction and develop customer-specific improvement measures from these.

A mix like this paves the way for a Zero Outage organization that works on the customer's specific problem, proactively tracks down systematic errors in the supply and service organization, and eliminates these quickly and safely.

Operational Quality: Zero Outage Ensures Reliability and Sustainability

Alongside an effective organizational structure, quality of ICT operations is also a cornerstone of customer satisfaction, and is necessary for successful long-term customer relationships. Not least because all complex systems – including information and communication technology – are also prone to faults at many different points. While fault avoidance will always fall short of 100 percent, any manager with ICT responsibility should attempt to hit this goal. Accordingly, in 2011 we at T-Systems introduced our integrated Zero Outage program, which impacts every stage of operations: before, during and after a potential fault. With clearly defined standards for platforms, processes and personnel, this program creates a basis for maximum availability and reliability. This is logical because standardization reduces complexity. This, in turn, is critical for avoiding faults or ensuring their rapid resolution. Fewer replacement parts, specialist expertise and internal experts with specific, individually held knowledge are required, and there are fewer unplanned effects when changes are made (see Kasulke 2013b). This minimizes IT outages and maintains the quality of ICT services at the highest level.

In the next chapter, we describe the key disciplines required for active quality management in ICT operations, and the specific approach we are following with Zero Outage. For **change management**, **incident management** and **problem management**, we have defined specific measures that produce tangible improvements in quality. This enables the preventive avoidance of faults, the rapid and structured resolution of outages and the permanent suppression of basic problems.

7.1 Process Fidelity as a Criterion for Success: Avoiding Outages by Attention to Detail

As in an audit, continuous measurement of process fidelity is important to analyze how often deviations occur compared to the target process – in change management, for example. In contrast to a simple check of results (such as the frequency of incidents), this clarifies whether a systematic and sustained level of quality has been achieved or whether the quality depends on the individuals involved – and whether changes in personnel could therefore lead to problems. **Process fidelity** is measured in terms of how often deviations occur compared with the total volume of core processes. Examples of this include: unauthorized changes, unprocessed problem tickets, escalations occurring too late in the event of a fault, or an increase in emergency changes. The last item in this list reveals that the control process is either usually started too late – and planning is therefore inadequate – or that there were errors in processing.

It is vital to look at the individual cases in detail to understand the background to process deviations and use these to derive improvements (training, process modifications, etc.). It has proved valuable to use a questionnaire listing the deviations as a checklist based on ITIL processes and company rules. This questionnaire should be used regularly in all customer accounts (see Kasulke 2014). We call this our Zero Outage compliance audit. A sample questionnaire of this kind is included in the Annex.

Standardization of Key Processes

Standardization of core processes creates a reproducible, uniform quality – that can be used globally – as well as straightforward instruments for control and management. This approach also prevents major deviations in service quality over time. A learning organization acts worldwide according to one standard in all core processes.

Standardized **global incident management** resolves acute errors as quickly as possible. It continually enhances its professionalism by repeating solution processes. The baton is then passed to problem management: it draws lessons from past errors and prepares a strategy to avoid recurrences at a global level in the future. Frequent sources of error are avoided by standardized change management. Each critical change is examined in a structured process (identical company-wide) and must satisfy the highest quality criteria, which are assessed and approved by the **Central Change Advisory Board (CCAB)**. The basis for process standardization is configuration management (CFM), which aims to supply current and consistent information about the configuration of the IT infrastructure. In the processes based on CFM (e.g., incident, problem, change and license management), decision-making is facilitated by the practical relevance and reliability of this information (see Kasulke 2013b).

Fire Drills

In incident management – and especially for top-priority incidents – managers should not simply put their faith in people learning from past mistakes and hope that the **alarm chain** will work as planned in the future. It is therefore a good idea to supplement the "stock-taking approach" with regular, unannounced fire drills (simulation of real incidents in order to ensure a seamless reaction when an incident actually occurs). These show if stakeholders stay SLA-compliant when responding to an incident. Downstream suppliers should also be included in such drills – not least because these suppliers are present as on-site incident support in the event of hardware failures or software configuration problems. The results of the fire drills are analyzed afterwards and discussed within the teams (see Kasulke 2014). In doing so, one interesting and unexpected result may be that the criticality of the customer's core business processes is not up-to-date and needs to be updated as a follow-on task, for example. In other cases, it might be discovered that the alarm chain agreed on paper with the supplier isn't effective in a real-world emergency.

CMDB Audits and Critical Landscape

An accurately maintained **configuration management database (CMDB)** forms the basis for almost any quality improvement. Changes to IT components must be entered by change management into the configuration management database. Ideally, this will be automatic. No **configuration item** should be added, modified, replaced or removed in the change management process without documentation. In complex back-end systems, it is difficult to represent the configuration items from various areas of IT in a single data model and a physical configuration management database. Typically, there will be multiple, specific configuration management databases, in which the configuration item data is maintained. A governing configuration management system (CMS) is then used to manage logical access to all configuration management databases and the information stored within them. This approach can also be used to map and visualize entire service chains.

For configuration management to be of a high quality, the information needs to be complete and correct, and the status of a configuration item needs to be understandable. This is an important point – since high data quality is vital for all downstream processes that will work with the data from the configuration management system.

From a process perspective, a well-maintained configuration management system also serves to unify a diverse landscape, and to establish key processes such as global patch and release management, for example. This means it is a critical factor in the avoidance of faults (see Kasulke 2013a).

The foundation for the CMDB is a customer-verified **critical landscape**, i.e., the systematic classification of all of the customer's applications, weighted by their importance for the customer's business. As the customer's business will undergo regular changes, the currency of the critical landscape should be examined and verified two to three times a year: this identifies the configuration items that should be classified as critical for an incident or change event. These items can then receive the consideration they need. The critical landscape is also useful for assessing how customer SLAs match the real world, and identifying coverage shortfalls or overlaps quickly enough (see Kasulke 2014). In reality, data quality is often inadequate, however, since the users of a CMDB (such as units involved in monitoring or incident/change management) are not always the same individuals that maintain it. In theory, the responsibility for CMDB maintenance is assigned to the operation manager or service chain operation manager. Routine audits of the data quality in the CMDB are therefore recommended, as well as random cross-checking of changes against the CMDB. The latter activity reveals whether the change manager has updated the affected configuration items and relationships.

7.2   Zero Outage Ring Fencing: Mastering Sensitive Situations with Maximum Vigilance

A few years ago, a public sector customer suffered a temporary service outage to its ICT systems. Police stations, hospitals and local government offices were unavailable for over six hours. Even senior officials were forced to use the mobile phone network. Within an hour, the outage was escalated to the Ministry of the affected German state. The CEO of the provider responsible was summoned to the state secretary's office and asked to explain when the problem would be resolved and what steps would be taken to avoid such a fault in the future.

How should such a situation be tackled? Apart from the obvious role of incident management, which will be discussed later, steps must first be taken internally to ensure that service provision is absolutely rock solid in the weeks after resolving the fault and to minimize the risk of additional faults. In situations with these kinds of requirements, the Zero Outage method uses a "ring fencing" technique.

The **ring fencing** in question is a package of measures designed to safeguard the most important applications at a customer over a short period of no more than a few weeks, so that faults can only occur in the most exceptional cases. Ring fencing is not merely deployed as a special safeguard following severe outages, but also for other important situations, such as a trade show presentation by the CEO, negotiations for complex extensions to contracts or in business situations that are especially critical for the customer concerned.

Measures include

- keeping changes – especially high-risk changes – to a minimum;
- additional manual monitoring of critical systems once an hour;
- hourly reporting to senior management;
- securing all technical resources and the availability of experts for all relevant topics on a 24/7 basis; and
- appointing a Manager on Duty, who personally oversees every change, every problem ticket and every incident, on a 24/7 basis.

Ring fencing is a resource-hungry method that is only deployed when the benefits can justify the costs. If it is maintained over a prolonged period, it tends to exhaust the organization and the effect – namely an increased sense of urgency – gradually ebbs away (see Kasulke 2014).

7.3   Global De-Escalation Management: Conquering Crises and Learning from Difficulties

De-escalation management is tasked with the rapid restoration of normal service after faults have occurred, and identifying the cause to initiate preventive actions for the future. De-escalation management comprises the Central Change Advisory Board, global incident management and central problem management. The incident management process aims to restore the normal level of service as quickly as possible, whereas the problem management process is designed to investigate the causes of faults and to implement prescribed solutions in order to avoid further problems – preferably on a permanent basis.

In general, the workflows for incident, problem and change management are based on the IT infrastructure library (ITIL) processes. ITIL is essentially a set of best practices for IT processes. It was first published in 1989 by the UK Office of Government Commerce (OGC) and has since undergone several significant revisions. In its current form, ITIL V3, the framework has many points of contact with other important standards such as ISO 20000, Six Sigma, COBIT and PRINCE2 (see Chapter 3).

To ensure a high level of quality 24/7 in the event of faults and to reduce the **mean time to repair (MTTR)**, the resolution of **major incidents** should be managed as a centralized process. In this approach, a core team develops routines for resolving highly critical faults, takes over the active management of suppliers in an emergency and develops recurrent solution models. The team focuses on the customer's critical business processes and provides high-priority support in this area.

One precondition for professional incident handling is that all information about the customer must be available. Only then can the real-world impact on the customer's business processes be identified in the event of a major incident. To this end, the critical landscape describes all of the key service chains.

To ensure that the incident management unit (potentially deployed worldwide) can perform its duties to the fullest extent, the team must be involved at the first signs of a critical impact. This also applies to potentially critical faults, whose classification (critical, high, etc.) has not yet been clarified. The earlier that incident management becomes involved, the sooner work can begin on resolving or avoiding the fault.

The most important factor for success in incident management consists of letting a **culture of urgency** flourish. Which is to say: Everyone involved does everything they can during an incident to restore the service as quickly as possible. Incident managers must be virtually immune to stress, must create order quickly and reliably in conflict situations, and be competent leaders of potentially globally distributed teams.

The central problem management unit then monitors root cause analysis after a major incident and implementation of the measures resulting from the fault. The unit must ensure that a problem encountered by one customer in one country leads to precautionary measures being taken for customers elsewhere in the IT service organization. In this way, the IT service provider's size offers a quality advantage for all customers (see Kasulke 2013a).

7.4 Global Change Management: Detailed Planning for Fault Avoidance

The next sections examine the central importance of change management to Zero Outage in detail. First and foremost, errors in **changes are the most common cause of a fault** – and should be avoided wherever possible.

IT systems are subject to continuous change. There are many reasons for this. The rapid pace of technical progress means users are continuously faced with new challenges in terms of speed and storage capacity. New kinds of applications are also made possible by new technologies. Requirements for mobile use and around-the-clock availability are also increasing, thanks to changes in our society. Existing IT systems must be serviced, and defective and outdated hardware must be replaced. Operating systems, firmware and software packages must be updated, and patched to fix errors and close security gaps. Newly acquired systems must be integrated into the existing IT environment, and legacy systems that are no longer in use must be decommissioned and disposed of.

All of this work must be planned and coordinated in detail, and a risk assessment must be performed for the consequences of these changes. This assessment must weigh up the likely impact on existing systems against the expected benefit these changes will bring. Steps must also be taken to coordinate the implementation schedule for these measures with everyone involved. Detailed plans must also be drawn up to allow for a degree of uncertainty and avoid interference from random factors. In the final analysis, the failure today of a system that was working yesterday can stem from only one of three causes: the system is being used differently (including faults caused by attacks or viruses), the system has a physical defect or – and this is easily the most common cause – the system has been changed.

This is where Zero Outage change management comes in. The idea is to systematically minimize risk when implementing changes to keep the level of disruption as low as possible.

7.4.1 A Practical Example for Change Management

Preparing for complex changes can be a very time-consuming activity. One example from the real world of business: a migration for a company running a frozen goods home delivery service. The only time slot available during the year to implement this migration was Easter, since no deliveries would be made for four consecutive days. Since any interruption or delay in the change process would have postponed the project by another year, planning and preparation had to be completed down to the very last detail.

So, what steps did we take for this mission-critical change? The most important step in change preparation is to consider all potential unexpected events and the appropriate response – i.e., what risks are associated with a change and how these can be prevented or their impact minimized. One technique Zero Outage change management uses here is to create a checklist for analyzing and evaluating the risks, which is based on experience gained from many thousands of changes and problems in the past. This list is used to develop the actual measures adopted in change planning and execution.

We submitted the change planning for technical appraisal by a small group of expert technicians. Then we sat down with experienced senior managers from IT operations and quality management to go through every step, to investigate every eventuality, and to review every possible interaction with activities running in parallel to the changes. In addition, our compilation of procedures for completing the changes – known as a "**runbook**" – was cross-checked by all of the suppliers and service providers involved. These checks included confirming the versions planned for software, operating systems and firmware; looking at dependencies on other software or hardware releases; and verifying that the overall implementation methodology complied with the manufacturer's best practice.

According to Zero Outage, testing is ascribed a crucial role in the planning and execution of a change. When conducted before execution, it can identify risks so that the corresponding countermeasures can be initiated. Testing can also be used to verify that the chosen approach only produces the consequences that have already been planned for. To achieve this, an adequate level of coverage must be ensured for the test method selected. The differences between the test environment and the production environment must also be analyzed and evaluated. Experience nonetheless shows that preparing for a change needs more than good-quality testing. It is at least equally important to review interim results and the modified system during change execution and – above all – once the change has been finalized. This ensures that the customer once again has full and unrestricted access to the system. In the case study mentioned, we performed these tests with the customer, so the customer could personally verify the positive result of the change.

As a result, we were able to perform the change for our frozen food delivery customer successfully, within the allotted time frame, and to the client's complete satisfaction. It goes without saying that outstanding results of this kind can be achieved only with an exceptionally qualified, experienced and highly motivated team. Everyone involved in the project fully **recognized the importance of risk prevention** and focused continuously on the Zero Outage goal of minimizing disruption to the customer's business processes.

This principle applies unreservedly to all stakeholders: to the Technician who performs a change, to the architects and operational managers who plan a change, to the review team that appraises the change, to the management staff that are involved in reviewing and approving such major and important changes, and – last but not least – to the customer-facing service managers who coordinate all of the key points in change planning and execution with the client.

For the majority of IT system changes, the criticality will not be as high as in this case study. Yet even minor changes – those day-to-day modifications in back-end systems that are regularly performed in large numbers – can have an impact that is just as great if something goes wrong during their execution.

7.4.2   Zero Outage: Ensuring the Success of Change Management

Zero Outage is supported by three key pillars: comprehensive quality assurance in planning, a high degree of standardization, and having safeguards in place during change execution.

What does this mean in detail?
Each change – i.e,. each permanent modification to a customer's IT environment – is assessed and evaluated with the same degree of diligence and care, and according to the same criteria. For this to happen, quality assurance must be systematically implemented throughout the entire company. A key instrument in this context is the Central Change Advisory Board (CCAB). Part of the global de-escalation management system, this unit reviews all important and critical changes within the IT landscape, and monitors their implementation. The CCAB sets standards for quality assurance and change completion, and monitors compliance with these standards. As a globally oriented unit, the CCAB can draw on a worldwide pool of policy competence. In an IT service organization based on cooperation across regional boundaries, local units need to work to the same set of international standards. This smooths the way for effective collaboration and ensures IT services can be operated for international customers at the same level of quality all across the world.

To ensure reviews of changes are as comprehensive as possible, the CCAB is supported in local units by satellite Change Advisory Boards (CABs). These work to the same standards and the same criteria as the CCAB, and review less critical changes occurring during the normal course of business. Know-how transfer between the CCAB and local CABs creates a global community whose work crosses national and organizational borders, and is usually performed "virtually", i.e., online. The community also approves any necessary changes to the official change management process, and implements this process worldwide. Since the CABs are tightly integrated with operational units, any changes needed to the process always relate directly to day-to-day operations in the IT environments.

In this way, all customers benefit from the experience of many thousands of changes per year. Common strategies developed for the execution of changes can therefore be deployed rapidly to any relevant areas. In the implementation of changes, highly standardized methods decisively reduce risks during execution and offer huge performance gains for planning efficiency. Tried and tested procedures, optimized planning with minimum downtime, optimum testability for results and detailed documentation of the individual steps within implementation ultimately work to maximize availability for the customer's IT services – or, put another way: Zero Outage.

So how do we put this standardization to work?
First, by copying the methods used for successfully executed changes into **change models**, i.e., creating templates for use with all future changes of this type. When planning a later change, the corresponding template is selected and then adjusted to the specific conditions of this new change. The degree of adjustment permissible is specified for each of the change models.

If this approach is systematically developed, it can lead to greater efficiency. If these changes always follow the same pattern within a precisely defined framework, the effort required in the **change review** – i.e., change planning quality assurance – is limited to reviewing the change models described above. In the planning phase of one of these model-based changes, all that needs to be done is to agree the timing with the customer and other operational teams to avoid conflicts with day-to-day business or any other change work planned.

Thanks to the high degree of standardization and the global application of these change models, improvements to the methodology can also be rolled out quickly worldwide. This increasingly anchors the Zero Outage philosophy in change planning. As regards the system architecture, outages and interruptions to system availability can be avoided by large-scale redundancy. This applies to data center infrastructure and network infrastructure, as well as to server hardware and to software design. If these requirements have been met in terms of the system architecture, change planning and execution can leverage this redundancy. In this approach, changes are implemented step-by-step on the redundant systems, so that services can be provided to the customer without any interruptions. By timing the change implementation to occur during low-traffic periods, the customer isn't even aware of any impact on system performance.

7.4.3   But What if Things Go Wrong Anyway?

What happens if the change does not proceed as intended? In the ideal scenario, the customer doesn't even notice it. To this end, change execution according to Zero Outage provides three key tools: detailed planning of the back-out method as an integral part of change execution, full testing of all steps, and the performance of critical tasks in change implementation in accordance with the **four-eyes principle**.

Here, it is crucially important to notice if things are going wrong while performing the change, and to respond appropriately. For each step in the change implementation, the expected result is defined as early as the change planning stage. During change execution, a check is made after each step to confirm that this result was obtained. If deviations occur, one of two courses of action can be chosen: either the safety margin is large enough to analyze the error and make a correction – or the change has to be rolled back. In the latter case, the detailed back-out method planned is used.

Minor deviations during change execution can be permitted by allowing a safety margin in scheduling. This can be used for analyzing and resolving errors, and in some circumstances the relevant suppliers may also become involved. To avoid lengthy lead times, supplier participation is secured well in advance for critical changes.

Planning safety margins of an adequate length will not be possible in every schedule. On other occasions, error analysis and correction work will simply be too extensive to be completed within the approved change planning window. In such cases, the changes made up to that point must be rolled back, so that the IT system is available in its starting state. The **back-out method** mentioned above is deployed in such situations. This method is drawn up during the preparatory phase for the change and is adapted specifically to the planned change implementation. Here, it is important that the methodology chosen for implementing the change doesn't cause data losses, even in the event of this fallback being used. With the Zero Outage approach, no change is performed without a back-out method being available.

Before finalizing the change, a comprehensive end-to-end test must be made to confirm that the customer once again has full access to the service. This test is run both when change execution is successful and in the event of a fallback being used. Where changes have a direct impact on the functionality of the entire system, the customer will preferably be involved in this test. Suppliers also participate in this end-to-end test. This validation of the change results is a standard within the Zero Outage program and is mandatory for all changes.

Service continuity planning also forms a part of critical changes. If it becomes clear during change execution that the change cannot be completed as planned or the impact on the customer service is greater than desired, the change is escalated to require the participation of the customer and internal management. The next step is the preparation of a contingency plan as a joint undertaking with the management team.

The analysis of incidents – i.e., the unplanned effects of changes – has resulted in the third tool in change implementation according to Zero Outage, namely the **four-eyes principle**. Previously, human error had occasionally resulted in errors being made during the implementation of a change. Even the most highly qualified and well-trained employees can still make mistakes. Technicians inadvertently shut down the wrong server, removed the wrong hard drive, disconnected the wrong network link or simply left out steps in the change workflow. In Zero Outage, the dual-control principle is a long-term strategy for avoiding such problems: each critical step in the change executed by the first technician is verified by an equally qualified second technician. This is an effective means of catching input errors, operating errors, typos and other simple human mistakes. While this increases the effort required to perform the change, this is more than compensated for by the prevention of incidents and their time-consuming troubleshooting.

7.4.4 Configuration Items: Down to the Very Last Detail

Let's now look at a more technical aspect of the Zero Outage change management process: configuration items (CIs). As we will see, CIs play an important role in the evaluation of changes. Put simply, each component in an IT environment is a CI. All of the servers, network components and software systems form part of the IT environment's configuration and are all CIs. While it is standard practice to create entries for these IT landscape components in a comprehensive database – accessible to all of the processes involved in rendering the service – Zero Outage actually goes one step further here. In Zero Outage, services are also CIs and stored in the database. In addition, each service is linked to all of its subordinate CIs in the database. With full details provided for each change about CIs that are directly or indirectly affected, one can accurately estimate the impact on the customer's business of executing this change.

The criticality of each CI for the customer's business activities is also noted in the database. Values range from "none" or "low" for test and development systems to "medium" for redundant systems and "high" for major business processes and, ultimately, "critical" for the customer's core business activities. If the impact of the change is determined for the corresponding CIs in change planning, this clarifies the effects or risks for the customer's business operations that will result from executing the change. This can be used to derive several strategies for later steps in change planning and implementation. Looking at the risks that have now been identified, is it advisable to execute the change in this format? To make this decision, the risk of not performing the change should also be established. If the risk is considered too high, possible decisions (potentially taken by mutual agreement with the customer) include going ahead with implementation as-is; identifying another, less risky mode of execution; or simply canceling the change entirely.

Another aspect that results from identifying the impact of the change on the corresponding CIs is the concept of classification. This is important for grading a change and the type of review a change is subjected to.

In the Zero Outage change management process, changes are placed in one of four classes: "standard", "minor", "significant" and "major". These classifications relate to the planned customer business impact (CBI), i.e., the analysis assesses the severity of the impact on the customer's business processes. With a "major" change, business-critical services remain unavailable to the customer for a specific period of time. With a "significant" change, business processes are only slightly affected. A "minor" change has no effect on business-critical processes or affects subordinate systems only, while "standard" changes have no impact at all on the customer's business processes.

In Zero Outage, "major" and "significant" changes are always reviewed by the CCAB mentioned above. Compliance with quality standards for "minor" changes is assured by the local CABs.

The planned impact on service availability for the customer is not the only criterion, however. The risk associated with the performance of the change must also be taken into consideration. Especially when an increasing number of changes are performed that do not cause direct disruption to server availability (by exploiting redundancies in the IT environment, for example), it becomes more and more important to keep an eye on the probability of a risk occurring – i.e., an unplanned interruption to the customer's business processes.

Accordingly, Zero Outage takes particular care to review especially high-risk changes, which it terms **special focus changes**. By working with the team responsible for operations, the architects, hand-picked specialists, the management team responsible, and service managers, the CCAB makes a detailed analysis of these changes. This analysis examines every aspect of the change procedure – including the planned and completed tests, the planned back-out method, the associated risks and the measures used to mitigate them, dependencies on other changes or other operations, and the necessary involvement of suppliers and the customer. The change is classified jointly, any weaknesses are resolved, and only then is the implementation approved. To keep all of the stakeholders informed of the current status and progress during the implementation of the change, each of those involved receives a report covering the key milestones.

For critical changes in particular, a long-term overview of planned changes is essential. This involves maintaining a change calendar to enable the early planning of change reviews and avoid potential conflicts with other changes. This also lets the IT service provider coordinate potential disruptions to service availability with the customer over a longer time frame.

Since the customer wishes to ensure reliable use of the IT systems and minimize disruptions to day-to-day business due to changes in the IT landscape, **official long-term maintenance windows** are agreed with the client. Depending on the criticality and change frequency for a system or service, these windows may be daily, weekly, monthly or annually. During these periods, which are planned with a long lead time, the customer has only restricted access to the IT systems. This plan gives the customer a point of orientation for business-critical operations, while it becomes easier for the service provider to plan the necessary IT changes.

Collectively, all of the measures described safeguard change management according to the Zero Outage principle. This minimizes the effects of necessary changes on the availability of the customer's IT systems and largely avoids having any periods in which the services are entirely unavailable. And in accordance with the Zero Outage principle, this applies not only to planned disruptions to IT system availability. By applying risk management systematically, unplanned disruptions can also be avoided while assuring compliance with predefined quality standards – and all whilst ensuring maximum standardization in the change execution process.

7.5 Global Incident Management: Prioritizing and Resolving Faults

To apply the standards of Zero Outage and process incidents with appropriate weighting (by severity from the perspective of the customer's business processes), our incident management can choose one of three priority levels.

The priority for an incident is derived from the "**CBI matrix**" (**CBI = customer business impact**, see Fig. 7.1).

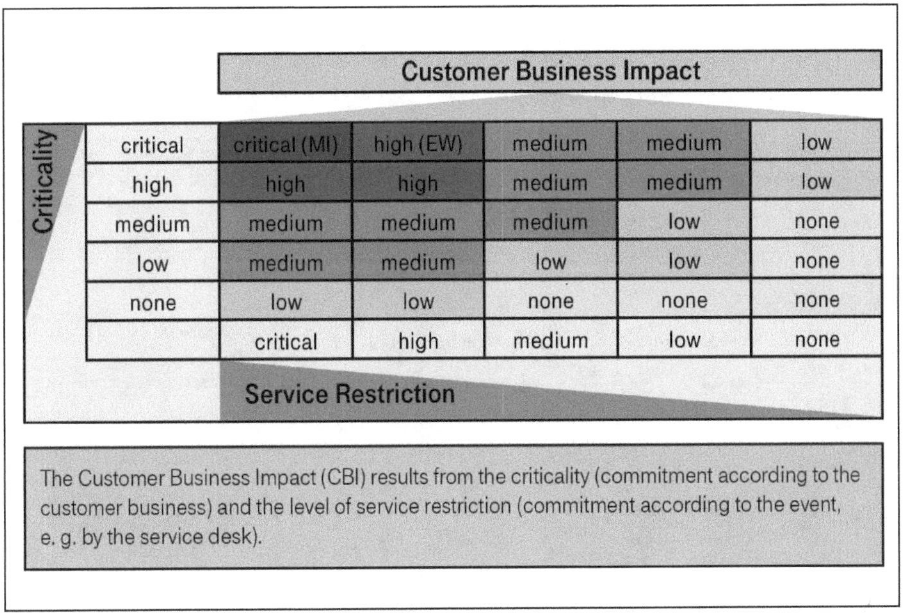

Criticality		Customer Business Impact				
critical		critical (MI)	high (EW)	medium	medium	low
high		high	high	medium	medium	low
medium		medium	medium	medium	low	none
low		medium	medium	low	low	none
none		low	low	none	none	none
		critical	high	medium	low	none
Service Restriction						

The Customer Business Impact (CBI) results from the criticality (commitment according to the customer business) and the level of service restriction (commitment according to the event, e. g. by the service desk).

Fig. 7.1 Standard Incident Prioritization – Assessment of CBI

CBI Critical:
One of the customer's critical service chains has failed completely, and there is no appropriate or economically justifiable workaround.

CBI High (EW = Early Warning):
Partial disruptions, performance problems or redundancy losses affecting one of the customer's critical service chains.

CBI High:
Partial disruptions or performance problems affecting one of the customer's non-critical services.

The CBI matrix determines (independently of tickets) the priority for the respective incident and must be established for every incident. With the CBI established, the correct incident procedure is initiated, as shown by the example in Fig. 7.2.

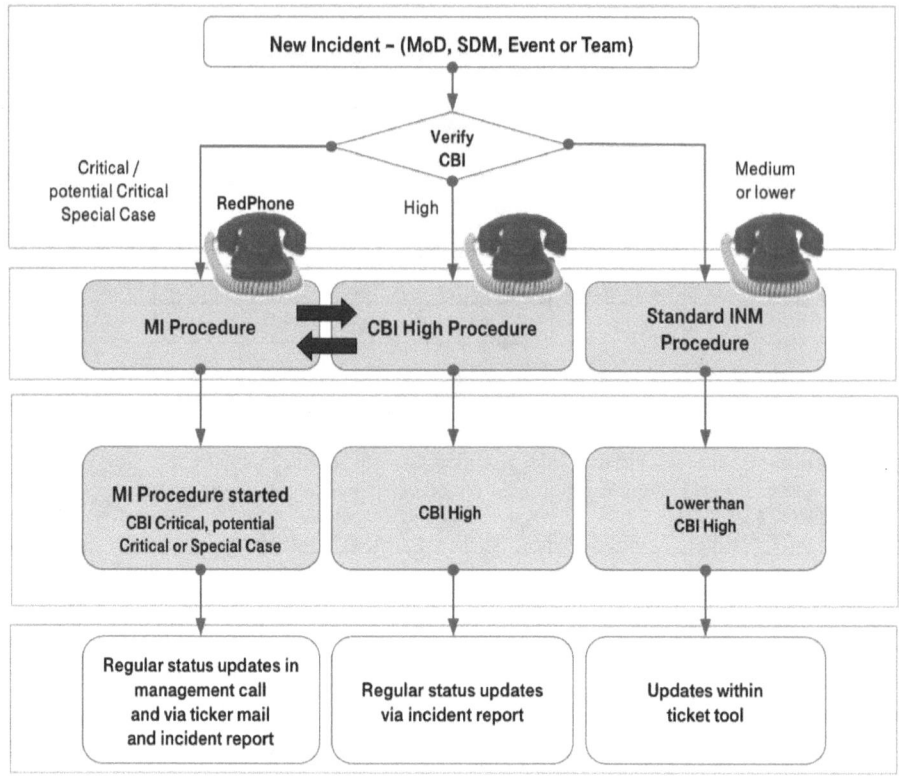

Fig. 7.2 Different Incident Procedures

Procedures in Incident Management

Standard Incident Procedure:
Standard incidents are processed within the responsible service units using the ticketing system.

CBI High Procedure:
The CBI High Procedure has the second-highest incident priority. Following verification of its CBI, this incident is passed by the technical knowledge worker directly to a lead incident manager (LIM), who then takes over its management with the participation of all necessary resources. The CBI is checked continuously by the lead incident manager as part of incident processing. If the CBI changes during processing, the LIM either escalates the incident to the **RedPhone** (see next section on tools) or de-escalates by initiating the standard incident procedure. The **lead incident manager** creates an incident report about the incident history and the actions taken, which is also used later when transferring the incident to problem management. After resolving the fault, the lead incident manager passes the incident to problem manage-

ment. In Zero Outage, the lead incident managers are assigned either directly to the customer – i.e., they are very familiar with the customer environment – or to the various production units, which means they are closely involved with technical aspects. In the event of a fault, both lead incident managers are involved.

Major Incident Procedure:
Major incidents have the highest priority for incident processing within Zero Outage. A major incident must be reported within 45 minutes of its CBI verification to the **RedPhone** staffed by worldwide incident management. This unit then initiates the major incident procedure **with the participation of all necessary Managers on Duty (MoDs) and suppliers.**

The major incident procedure follows a chain of predefined process steps that are familiar to all participants in the Zero Outage program. All of those with incident management responsibilities complete regular training in the major incident procedure. Knowledge of the procedure is also tested by conducting random spot checks. The major incident procedure can be triggered by a MoD, a service delivery manager (SDM) or by a LIM contacting the RedPhone. The actions triggered are measured in KPIs and are subject to continuous improvement.

Key Characteristics of the Major Incident Procedure Include:
- Management by the RedPhone as the central instance with trained personnel, providing appropriate expertise for 24/7 support of incidents in the highest category
- Direct involvement of partners and suppliers in conference calls to ensure maximum leverage of manufacturer expertise
- Checking changes from the last seven days as a mandatory task
- **Full layer check** (a check of all CIs and components of the affected service from the network to the application) utilizing predefined checklists and instructions to ensure that all technical aspects have been considered
- Involvement of senior management and the MoD in the major incident procedure
- Continuous customer communication
- Regular updates concerning all actions performed and their results

The diagram in Fig. 7.3 provides an overview of the overall major incident procedure.

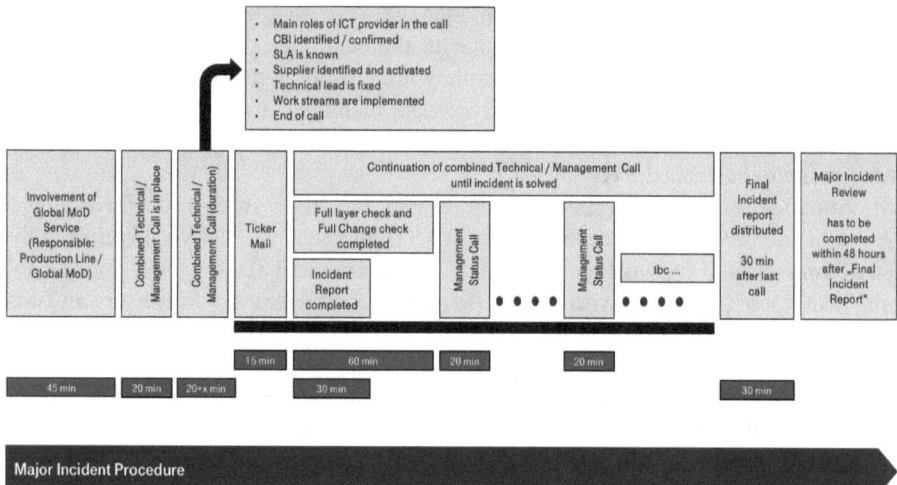

Fig. 7.3 Overview of a Major Incident Procedure

Implementation Tools

The tools essential for implementing the process are the RedPhone as a global instance, supplemented by clearly-written process descriptions in the form of "CookBooks".

RedPhone

The RedPhone is available to all country organizations and organizational units that handle the processing of potentially critical and critical incidents. It initiates the major incident procedure and manages the entire process from the deployment of necessary resources to alerting the executive management team. In metaphorical terms, it is the ICT provider's "fire brigade service" that is there to tackle fires at the customer. Staff work in shifts to ensure that the service is available 24 hours a day, seven days a week.

The RedPhone comprises **global incident control** and **global lead incident management**.

Global incident control is the **single point of contact (SPOC)** for the major incident procedure and has the following duties:

- Operates as a single point of contact for the major incident procedure
- Initiates joint conference calls with technicians and managers 20 minutes after being alerted
- Personally invites participants according to the MoD structure
- Provides an end-to-end communications infrastructure (conference calls, desktop sharing)
- Acts as the central information clearing house: all major incident information is sent to the RedPhone as the central instance, from where it is distributed to all stakeholders
- Includes partners and suppliers according to the Zero Outage program

- Completes documentation (event triggers)
- Sends reports (Incident Report)
- Alerts the executive management team

Global lead incident management is responsible for **managing the structure and content of management conference calls**. Calls are hosted by a global lead incident manager, who shepherds the incident until its resolution. This person must ensure that the processes follow a pre-defined structure, which involves the following tasks:

- Determining the customer business impact (CBI) for the customer
- Checking whether a workaround is possible
- Checking changes from the last seven days
- Initiating a full layer check
- Requesting the participation of partners and suppliers in finding solutions
- Regularly updating stakeholders via a short summary ticker mail
- Requesting safeguarding measures once the incident has been resolved
- Preparing the minutes from steering calls

The global lead incident manager uses regular follow-up conference calls with management participation to ensure that analyses are performed continuously, and requests regular updates on developments in accordance with the predefined structure. Once the major incident procedure is complete, he or she is also responsible for the handover to problem management.

CookBooks

All of the steps described for the major incident procedure and CBI High procedure are explained in "CookBooks": process descriptions that are written in simple and straightforward language. CookBooks exist for all core processes (incident/problem/change/configuration management) and are regularly updated to reflect new information. With their straightforward language, CookBooks help employees to understand procedures quickly and apply them in day-to-day practice.

Yet the most important aspect of Zero Outage incident management is the attitude to one's opposite number taken by all stakeholders throughout the organization. This includes a well-developed sense of urgency, the tracking of multiple solution approaches in parallel (plan B, plan C, etc.), the personal involvement of top management and the unwavering focus of all participants on taking any action whatsoever at any point in time to resolve the fault as quickly as possible.

War Room

In special situations, such as an increased frequency of serious faults within a short time at a single customer, the establishment of a permanently-staffed "War Room" has proven its worth as a strategy. The War Room coordinates the continuous monitoring of critical systems and alerts all of the employees necessary for resolving faults in an emergency. A War Room should have direct authority from top management to take immediate and direct action to guarantee operational stability.

7.6 Global Problem Management: Learning from the Past, Optimizing for the Future

In the Zero Outage program run by T-Systems, problem management (PRM) is tasked with preventing the reoccurrence of incidents and thereby reducing their quantity. Another aspect is determining the potential for improvements from faults – in relation to platforms, people and processes. Problem management not only discovers the reasons (**root causes**) for faults but goes beyond this to determine all of the weaknesses that could adversely affect the services provided to the customer. For all of the potential improvements, measures and solutions are defined and implemented. Problem management is initiated both after faults have occurred (reactive PRM) or before faults occur (proactive PRM) on the basis of identified patterns, complaints, warnings and knowledge gained from earlier faults at other customers or on other platforms.

Reactive problem management is typically triggered once the fault has been resolved or the availability of a stable workaround has been confirmed. In contrast to incident management, which is active 24/7, problem management does not involve shift work, and case handling does not use a "follow the sun" approach to establish a permanently staffed team for performing root cause analysis. Instead, cases are handed over in conference calls at pre-defined times: incident management passes the cases on to problem management with all of the relevant facts and figures. For the most important, critical cases, an incident review is performed at the end of the incident process. This review is intended to finalize documentation, list any pending questions and appoint any individuals who need to cooperate with problem management. Where possible, the assigned lead problem manager is invited to attend these reviews, so that he or she can receive information first hand and raise any points requiring clarification.

For other cases, the Zero Outage problem management community organizes handover to the responsible problem managers, who are then provided with the necessary information. This also takes place at pre-defined times. Participation in these conferences is mandatory.

If, in exceptional circumstances, a stable condition cannot be established during the incident and there is a high risk of reoccurrence, the cause of the fault is investigated by the lead incident managers in incident mode.

7.6.1 A 360-Degree View: Analysis of Root Cause, Workflow and Consequences

In Zero Outage, incidents are graded according to the customer business impact (CBI), which is defined by the service's criticality for the customer and the extent of the potential damage (see the CBI matrix in incident management). If the reason for an incident is not discovered in the incident process, problem tickets are created and the technical cause is then investigated.

For faults graded at the two highest CBI levels (critical and high), a **comprehensive root cause analysis** is conducted, which goes far beyond the simple technical aspects of the platforms and includes examining the associated processes, and the actions taken by internal and external personnel involved in the fault. This analysis is supported by an **extensive questionnaire** that serves as a checklist for the responsible lead problem manager and his or her appointed team.

The aim here is to identify the exact sequence of events leading to the fault and its resolution. Were there delays in the alarm chain – if yes, why? Were all necessary resources available? Did all persons (both in-house and at the supplier) capable of contributing to the solution participate promptly in the conference calls? Did the persons involved have the right expertise? Did our own monitoring systems trigger – or were we first alerted by the customer? Was best practice followed for fault clearance? Were all necessary checklists available? Were there any problems with customer communication? Was the impact on customer business clear – and therefore the CBI? Were the service and its criticality known and correctly defined and described in the configuration data? Was the service a new service – only recently introduced by a project, and for which service readiness had perhaps not yet been established?

In the review process, the legal situation must also be considered. Were there breaches of **service level agreements (SLAs)** or **operational level agreements (OLAs)**? Did suppliers respond promptly and fulfill their contractual obligations? Do contractual penalties or claims for compensation apply? Does the service purchased meet the expectations or needs of the customer? Answers to these questions are needed in the first place to facilitate customer communication. If SLAs were not breached but the CBI is nonetheless critical, these answers will also be needed to amend the SLA concerned.

The question of who is responsible for the fault – the provider, the customer or one of the customer's or provider's suppliers – is generally important for discussing potential penalties. In Zero Outage, however, this aspect is also part of a comprehensive and integrated assessment of all risks at all stakeholders.

7.6.2 Root Cause Analysis: System Failure or Human Error?

One of the most common causes of incidents is a change. If this is the case, a detailed investigation is made of the responsible change, in cooperation with change management. The most important questions to be clarified here include: was it due to change planning or change execution? Were tests adequate beforehand? Did the change pass through the change management process properly, and was it approved by the appointed committees? Did suppliers also check and approve the change as necessary? Was the change runbook complete and correct?

Only in a handful of cases is the cause of incidents to be found exclusively in a system failure. In previous chapters, we used the example of a fault resulting from a rodent gnawing through a cable to illustrate how multiple cases of human error are typically the reason why systems or interfaces with redundant safeguards can be disrupted. This is why Zero Outage also asks the "painful" questions: was reckless-ness or ignorance actually the reason for the error? Were processes being tracked? Were there gaps in the process or definition? Was there a lack of know-how? Were actions taken only by the persons authorized to do so? Was the four-eyes principle applied, if defined for the activity?

Technical root causes produce errors in hardware, software or configurations. The task here is to unpick an often-convoluted network of inter-relationships. Since complex environments almost always involve suppliers and their components, sup-pliers are also included and asked to provide help in completing the analysis. With technical defects, questions must also be asked about why the redundancy set up for potentially critical services failed to work. To identify the underlying cause, the principle of the "Five Whys" must be applied, i.e., "Why" questions must be asked until the cause has been clearly established. To illustrate this, we can use the ex-ample of an email service that was interrupted due to a fault. By applying the prin-ciple of the five "Why" questions, this iterative approach to problem management discovers that a) network communications were disrupted; b) there was a firmware bug in the firewall; c) the firewall was running outdated firmware; and d) – the actual cause of the fault – there was a problem in patch and release management. To avoid further faults of this type, appropriate strategies must now focus on resolv-ing the last of these issues. As this example makes clear, the principle ensures that investigations do not simply stop at the level of symptoms, but "drill down" further and further until the underlying cause of the problem has been identified. It may also be necessary to replicate specific environments in labs in-house or at suppliers, or dive deeper into the problem by making detailed inspections of software and hardware components.

Only when the repercussions of a fault have been correctly and clearly classified can the appropriate activities follow. A number of questions need to be clarified: how long was the service unavailable or only partially available to the customer? Was the failure total or did it only cause restrictions? Was a workaround available to the customer, so that business processes were able to continue? What losses were suf-fered by the customer? If faults involve platforms used by multiple customers, these questions need to be answered on a per-customer basis. If the fault could affect other customers and platforms, then precautionary steps must be taken in order to prevent an occurrence in these other locations. Apart from identifying the causes of faults and weaknesses, the successful aspects are also documented to record both the positive and negative insights to share with all participants in "lessons learned" ses-sions.

7.6.3 Post-analysis: Setting-up Measures for Long-term Deployment

Once the causes and weaknesses have been found, measures are introduced for ev-erything that resulted in problems. On the one hand, this prevents the specific fault from reappearing. On the other hand, it ensures that the potential improvements identified are utilized to improve the company in terms of its people, processes and platforms, and reduce the overall number of faults that occur. Before a measure is introduced, a precise definition is made of the individual steps to be taken. Any deliverables that need to be produced – such as new systems or ICT services for the customer – are also specified. Each measure must also be assessed to see if it applies to the current case only or could be used proactively as a "multiple measure" to protect other at-risk customers or platforms from faults. For each measure, an owner is defined who has responsibility for the measure until its closure.

Measures can be technical topics that are implemented during the change process. This can quickly assume the proportions of a project – if large patch lists are created and need to be worked through, for example. Training courses or other activities involving personnel can assist to reduce human error. If there are deficits in the services sold, then sales activities can also be included in these measures. If weaknesses have been identified in the process, process improvements need to be implemented. Customers or suppliers may also become involved if necessary – i.e., if they are found to be the source of root causes or weaknesses. Another important aspect here is the implementation of **fault simulations** as measures for specific weak points (alarm chain failed, redundancy was ineffective, etc.). These **fire drills** can be used to test processes or the functional capability of systems, or exercise the skills of personnel who are involved in the process.

As improvements are put into practice, they should be monitored closely. Adherence to deadlines should be tracked, for example, and reviews performed after the implementation of especially important measures. In this way, the full and successful implementation of all measures can be verified.

7.6.4 Stakeholders in Problem Management: Everyone in the Same Boat

A successful root cause analysis requires full participation in the problem management conference calls. The lead problem manager steers and documents the analysis, organizes calls as often as needed and defines the tasks that each of the participants must complete. Typically, lead incident managers provide problem management with their information by means of the incident documentation and the incident review as mentioned above. As the customer-facing interface, the service delivery manager (SDM) obtains the necessary information about the impact of the fault directly from the client. Typically, the SDM also handles measures related to sales and informs the customer of tasks they have been assigned. To facilitate the frank discussion of weaknesses internally, direct customer participation in problem management conferences is reserved for exceptional cases. The SDM presents the customer with the results of the analysis.

The experts consulted perform the technical and process-related investigation work and submit their findings. If a change resulted in an incident, the responsible change manager is also consulted in order to provide full details of the change and implement any measures for improvement required. Employees at the supplier can also be involved if necessary.

Once all of the results have been processed and documented, the lead problem manager asks each of the participants to confirm the analysis and the measures as defined. The way is now clear for the final step in the process: the analysis is presented to the relevant management committee for acceptance.

7.6.5   Involvement of Partners and Suppliers

Since vertical integration is declining at many partners and suppliers, it is often necessary to include them in the root cause analysis. To bring these people on board quickly and avoid the need for quotations to be prepared beforehand – which would result in delays – **Zero Outage agreements** have been concluded with key partners (see also Chapter 14, supplier management). These agreements ensure that key service providers are available for the problem management process as soon as possible and can complete their share of the root cause analysis within a specific time frame. Collaboration with partners is evaluated later in the problem management process and the results are provided to strategic partner management. These results are also used as input for the service calls held at regular intervals to benchmark cooperation with partners against KPIs and optimize where necessary.

If it transpires that the customer or supplier bears responsibility for a fault, then problem management is assigned the task of identifying the specific errors made by the customer (e.g., a failure to approve a change that could have prevented the incident) or the supplier (e.g., defective components), and estimating the effort that was required to resolve and process these failures or defects. This data is then used as input for claim management. In terms of customer responsibility, this task is assigned to service delivery management, and in terms of supplier responsibility, this is a task assigned to supplier management working with procurement. These units ultimately decide whether or not a claim is then opened and requested.

7.6.6   No Problem Management Process without Professional Closure

To ensure that top management is also involved in problem management, **sign-off calls** are performed for critical and highly visible incidents. The relevant top management levels from the affected operations units, service management and quality assurance are invited to attend these calls. During the call, the responsible lead problem manager presents the cause and requests acceptance. This helps to inform the company executive about existing faults and their causes. In addition, the calls are also a forum for offering additional helpful advice and clarifying controversial issues.

A similar principle is also at work with the **quality gate**, which is convened at the next level of management for all cases with the second-highest criticality. Here, a joint decision is made about accepting the analysis and whether conditions relating to re-work will be specified. Active participation by senior management is an important factor for success, not least because this defines the attitude taken to errors: open and direct, solution-oriented – or merely superficial.

For documenting the root cause analysis, a standard Zero Outage **questionnaire** is used. The responsible problem manager can simply press a key to import the data from the incident report, and use this data to work through the subsequent questions and document the measures. This document can then be used to generate the databases for high-level analysis work, for measure tracking, for entry into the Known Error Database and for reporting.

For sign-off calls and special presentations, the findings are summarized in a "root cause card" in PowerPoint format.

The root cause analyses are used to fill the **Known Error Database**. For each case, special keywords, symptoms, solutions and customers are entered. This serves as input for the incident managers, giving them the opportunity to search for known errors in order to accelerate the incident process.

7.6.7 Global Problem Management in Zero Outage: The Four-Eyes Principle

To professionalize problem management and ensure a consistent process, problem management has been established as a centralized, global unit in Zero Outage. Directed by a global head of problem management, the unit's other members include global process managers and a small group of experienced lead problem managers. This unit is assigned process and operational responsibility for problem management in Zero Outage. The unit's work includes defining methods and processes for problem management, and maintaining the master problem management CookBook, which documents the official company-wide process. The unit also manages the annual improvement program for problem management and handles the performance of root cause analysis work for major incidents. Other responsibilities include onboarding other problem managers in the company and providing training as needed, as well as hosting "lessons learned" conference calls to which all problem managers worldwide are invited. These calls are used to discuss successful and problematic cases to suggest improvements or ideas and to share insights.

Problem managers have also been appointed for major customers, national companies and a number of company divisions. These individuals report to the global problem management unit and perform root cause analysis within their areas of responsibility.

A weekly **process improvement team (PIT) call** is also held with the problem managers in the company divisions, to discuss high-level operational topics and take decisions on process improvements. This approach fosters the propagation of good practice in the wider organization.

As there is often a lack of experienced problem managers in the overall organization, quality assurance in Zero Outage is organized as a core function. The first step is applying the four-eyes principle to audit the root cause analyses of the first two criticality levels. Any deficits are highlighted in a written report that is sent to the authors of the analyses. The analyses are accepted only when the desired level of quality has been achieved, and can then be considered complete. A pre-defined list of criteria is used here, with only about 10 percent focusing on formal aspects while the remaining 90 percent address the analysis content.

In addition, a check is made once a week to confirm that all incidents from the two highest criticalities have already been processed within problem management. If this is not the case, the pending incidents are communicated to the responsible individuals, who are then assigned the task of analyzing these cases. This ensures nothing gets "swept under the carpet."

7.6.8 Proactive Problem Management

Alongside reactive problem management, which focuses primarily on the need to implement measures for other customers or platforms to avoid the same faults, Zero Outage also utilizes proactive procedures that are designed to prevent the occurrence of similar faults. Within the context of Zero Outage, an analysis of all incidents is completed once a quarter to identify recurring patterns and establish multiple measures. Another element covered in all root cause analysis work is categorization by cause type and the weaknesses identified. The tools used are oriented on less critical faults, such as those illustrated in Fig. 7.4.

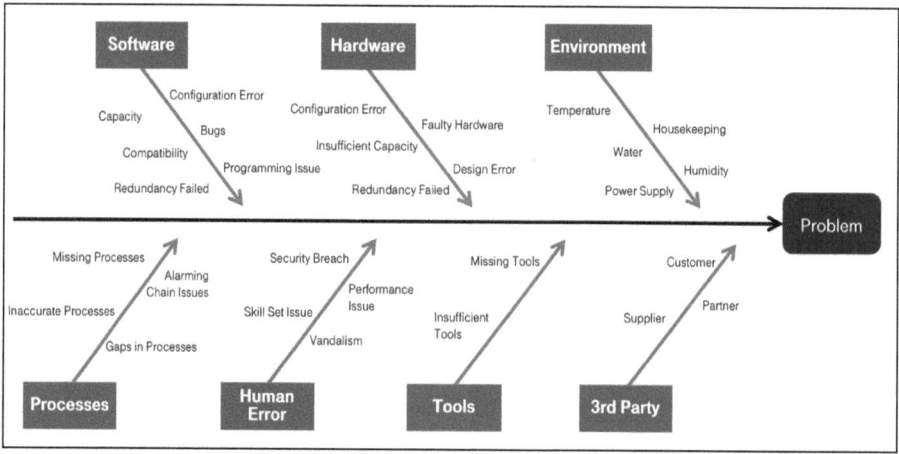

Fig. 7.4 Root Cause Categorization (extract)

These identify similar or frequently recurring incidents, and then apply a common root cause analysis to all of these incidents to prevent escalation into high or critical incidents. Configuration items/equipment where errors or warnings occur more frequently are also investigated, and then prioritized for monitoring.

The next step in developing this process further will be to use Big Data to perform additional, more flexible analyses across the entire dataset. As has been shown above, problem management forms the basis for the Zero Outage Quality Roadmap, which will be described in detail in Chapter 8, and thus for all systematic quality initiatives.

7.6.9 Integration with Incident and Change Management

Another important factor for the success of problem management is its close integration with neighboring processes such as incident management and change management. Cases are passed between incident management and problem management on a daily basis. Regular meetings are also held to identify positive and negative aspects of case handling. Analyses of interest are presented to incident management, and incident and problem management also cooperate closely on the incident review – or "warm handover" – used for critical cases. Change and problem management also work closely together on faults that have resulted from changes. Technical measures that have been defined in problem management must also be properly implemented in the change process.

7.6.10 Outlook: The Future of Zero Outage Problem Management

The first place where the potential for optimizing problem management can be found is in the proactive methods, where Big Data can offer a decisive advantage. Training and coaching problem managers can also improve the level of quality in analyses, producing measures for optimization that are even more effective and precise. In terms of content, the contractual issues arising between the customer, provider and supplier should be one area of particular focus. Another aspect that has proved to be important, particularly for new services, is greater participation in migration projects and transition/transformation situations to help ensure that more attention is paid to all operational aspects at an early stage.

7.7 Operational KPIs: Making Quality Measurable

If measurements aren't taken, improvements can't be made. Weekly reviews of quality KPIs (key performance indicators) by top management help to ensure that the pursuit of ever-increasing quality is also a point of continual focus in day-to-day business practice. Top management is not only informed about critical operational faults and projects, but also about measures for improvement intended to provide long-term solutions to quality problems.

To measure quality, a number of key figures can be defined to accompany the core processes of incident, problem and change management.

7.7.1 KPIs in Incident Management

In the field of incident management, two primary KPIs are applied for the measurement of operational stability. The most visible of these KPIs is the number of major incidents (MIs).

To be able to classify faults correctly, it is important to document the customer's "critical landscape" and to compare this documentation with contractual agreements. In the event of a fault, the restriction on the affected IT service must also be verified in consultation with the customer.

A short-term positive effect on the number of major incidents can be achieved by first analyzing the most common causes of errors and responsibilities. These calculations are based on data from faults in critical IT services that resulted in a total failure (MI) or a partial outage with the risk of a complete failure (Early Warnings or EWs for short).

Note that the **three primary causes of major incidents** are:

a. Fault lies with the customer: the customer or a customer supplier either caused the outage or has responsibility for the affected IT system.
b. Fault lies with a supplier to the IT service organization: outages of this kind can be shortened by setting up common quality initiatives or defining rapid escalation pathways.
c. Fault lies with the IT service organization: outages in this category can be directly affected by the provider's own quality initiatives.

If the IT service provider is serious about achieving high quality over the long term, then all three sources of error must be addressed – even those not under the provider's direct control.

The other KPI of importance in the field of incident management is the mean time to repair (MTTR). The mean time to repair is a useful quality benchmark for incident management and can highlight aspects such as the real-world effectiveness of the alarm chain or the speed with which the necessary qualified employees were made available to resolve the fault. Regular spot checks ("fire drills") and a focus on organizational quality can improve the mean time to repair, even over the short term (see Kasulke 2013a).

7.7.2 KPIs in Problem Management

Building on the work of incident management, problem management has the aim of investigating **the causes of faults that occur** in detail, and defining **measures for their long-term resolution**. There is a qualitative aspect, i.e., ensuring that root cause analyses are fully completed to leverage the maximum potential for improvement. Time is another aspect considered to ensure that the results are made available as soon as possible. This works to minimize the risks and ensure early communication with the customer.

The most important KPIs in problem management are:

Root Cause Quality: a list of criteria is made available to measure the completeness of a root cause – i.e., the correct completion of mandatory fields, for example, as well as the clarity and scope of the root cause description provided. For the analysis to be accepted, a coverage of at least 95 percent must be achieved. A formal review is no substitute for an in-depth analysis of the content, however. Accordingly, care must be taken to ensure that all of the fault's causes and contributory factors have truly been understood.

Average Time for Root Cause Finalization: for each fault analyzed, the average time taken for a complete analysis, including a definition of the measures, must not exceed 15 days.

Preliminary Root Cause Rate: an initial root cause should be made available for 60 percent of cases within the first three working days. A full root cause analysis is not required. But if key reasons have been discovered, they can already be communicated to management and the customer.

Measures Rate in Time: at least 90 percent of the measures defined must be implemented by the agreed deadline. This rule ensures that measures are not simply "put on ice".

Root Cause Found: measures the percentage of cases for which the root cause is found. This figure should exceed 90 percent.

Root Cause Rate in Time: specifies whether the cause for a fault was found within the time prescribed and as agreed with the customer (a three-day window, for example).

Problem Management Solution Rate in Time: specifies which percentage of countermeasures to outages have been implemented successfully within their planned implementation periods (see Kasulke 2013a).

7.7.3 KPIs in Change Management

The change management process focuses on the active management – i.e., the addition, changing or resolution – of elements within the IT infrastructure in accordance with standardized methods and procedures. The following KPIs are applied to measure the quality of these changes.

The number of major incidents caused by changes indicates how many critical outages occurred despite dedicated change management. To learn from the causes of failed changes, data needs to be analyzed in close consultation with problem management.

Other KPIs used to measure the quality of change preparation and execution include the "Ratio of Successful Changes" and the "Ratio of Changes in Time". The successful implementation of changes without errors must also be supported by not exceeding planned downtime for the affected IT systems while keeping these windows as short as possible. Therefore, the KPI "Ratio of Changes in Time" is measured. This figure specifies the proportion of the total change volume that was implemented within the planned change window (see Kasulke 2013a).

The Zero Outage Quality Roadmap: Ensuring Quality Step by Step

Now we come to one of the most important elements of Zero Outage: the central Quality Roadmap. This is the "master plan" for systematically and effectively managing operational risks in critical customer systems over the course of years. The Zero Outage Quality Roadmap ensures that, instead of getting lost in individual problems, all risks that lead to operational disruptions are **recorded in a structured way, regularly and reliably assessed and, finally, resolved according to their priority**.

What does a Zero Outage Quality Roadmap look like, and how can one be created? The answer is simple and follows the logic of ITIL problem management: learn from mistakes.

We can also look to other – older, more mature – industries here. In the aviation industry, every accident is investigated thoroughly. The findings are documented, and the investigators determine whether the accident means that structural changes have to be made to the aircraft, whether flight procedures (checklists) have to be expanded or modified, or whether pilots may need additional training. There are also reliable procedures for notifying everyone involved about these changes – such as other pilots, air traffic control, aircraft manufacturers, trainers or aviation authorities.

This is exactly how the Zero Outage Quality Roadmap works. First, the major disruptions over a specific period of time – we recommend a year – must be thoroughly analyzed. It is important to be ruthless in this and make no allowances for internal or external sensitivities. The aspect of **"ruthless analysis"** will be illustrated in the following using an example from the starting phase of the first Zero Outage Quality Roadmap.

Let's look again at the example from Chapter 2, in which a data center site experienced a severe operational disruption when a fault arose in the connection between two data centers. This meant that data could no longer be synchronized between them, leading to the outage of various applications for several important customers. It turned out that the cable had probably been gnawed through by a rodent.

How was this outage incorporated into the Quality Roadmap? Firstly, we are dealing here with an obvious risk that has to be included in the Quality Roadmap: the risk of interruptions to physical WAN connections that have been chewed by animals. Could this have been prevented? Yes! It was common knowledge that the cable crossed a river and was thus at risk of rodent bites. A steel covering on the cable could have prevented this disruption.

But this was only the superficial cause of the outage. To include every single risk in the Quality Roadmap, we have to know every reason for this outage. The key question is: what could have been done to prevent it?

The data center connection was configured redundantly, of course, so another obvious cause of the incident was a non-functioning failover, meaning the seamless switchover to the secondary connection. Further analysis revealed an error in the firmware on a router that prevented the smooth failover. So, the second reason for the outage was a software error from a router manufacturer.

But we are not even close to being done, because this was actually a well-known firmware error from the hardware manufacturer. The third reason, therefore, was in-adequate patch and release management. The affected operating unit should have known about the new firmware and installed the latest release long before the incident.

Another question is why the failover functionality was never tested. This means that the fourth reason for the outage was a lack of adequate testing after the latest firmware was installed. And reason number five was that apparently no regular tests were carried out to ensure a functioning failover.

The incident occurred at 2:30 a.m. Although the monitoring system had identified the incident, troubleshooting didn't start until 5:30 a.m. Response was delayed be-cause the system had incorrectly assessed the event's criticality and thus failed to trigger the chain of alerts that would ensure every important technician and partner was working on the problem as quickly as possible. Had everyone been available immediately, the incident could have been resolved before the critical systems went into operation. This was reason number six: weaknesses in the monitoring system and chain of alerts.

We can see from this incident (which happened exactly as described) that several factors have to coincide to trigger a catastrophe in a high availability system. This is where we can see the opportunities and advantages offered by systematically inves-tigating incidents based on a Zero Outage Quality Roadmap. A careful analysis of every incident can teach us a great deal if every factor that has led to the incident is included on the list of risks. When a large number of incidents are analyzed, the input for the Quality Roadmap becomes a list of all risks that could lead to operational disruptions, as well as the frequency with which these risks have been encountered. After Fukoshima, where things also happened for the first time, we supplemented this approach with "paranoid brainstorming," which involved coming together as a team and adding even more risks that had not arisen but could possibly be encoun-tered at some point. One new risk factor that we, for example, took into account was an employee strike.

The **risk list** makes it possible to define corresponding **countermeasures**. You can then establish a program for the coming year by taking into account the frequency of an incident and the cost of systematically resolving the problem for every impor-tant customer and system. This program first eliminates the most important sources of error for all customers – but without forgetting the other risks that were identified but not currently considered critical.

This Quality Roadmap is expanded and updated each year to include new problems, and we work through it continually, systematically and comprehensively in the interests of Zero Outage.

We have always divided these programs and measures into the previously discussed categories – **people, processes, and platforms & technology** – by asking the following questions: what courses of action need to change? What can be done in terms of technologies and procedures to prevent such outages?

If priorities change suddenly, a previously defined part of the Quality Roadmap could be implemented immediately. The risk of an employee strike, for example, may be considered low for a long period of time, but it can quickly become a pressing matter in the event of spontaneous political changes.

Of course, it is usually not possible to completely eliminate a cause of error within a year. Using what is probably the most frequent source of error as an example – namely, an "incident caused by a change" – we want to describe how we have repeatedly introduced new risk minimization measures in the Zero Outage program.

In the first year of Zero Outage, we started by standardizing the process globally. In doing so, it was important to keep the approach simple, practical and understandable, and to carry out audits in every unit and country to ensure that the process was being followed and implemented as defined. We developed a Zero Outage process compliance system for this, which compared the behavior of the operational teams with the target process in incident, change and problem management. Deviations were identified, measures were introduced and new audits were then conducted. One central component of this process is a standardized risk assessment for all changes in order to achieve a binding risk evaluation. In this first phase, we also defined KPIs and took regular value measurements – following the adage of "if you can't measure it, you can't improve it." We thus determined the base values in each operational unit with a focus on Zero Outage in change management. The most important KPI here was "incidents caused by a change."

In the second year, the most significant process change was the establishment of a Central Change Advisory Board. This has resulted in two key changes: a central team now monitors particularly difficult, complex or otherwise risky change processes above a certain risk classification, and additional backup procedures are carried out together with the manufacturers of the respective hardware and software components. The implementation of the changes is monitored as well. These measures reduced the incidents after a change by 48 percent in just one year.

Nonetheless, according to the Quality Roadmap, changes are still the most common cause of operational disruptions. For this reason, additional measures were initiated in the following years. First, a standardized and mandatory training program with annual recertification was launched for all employees, and the four-eyes-principle for checking and implementing changes was also made obligatory for the next-lower risk class. This resulted in a 58-percent decline in incidents caused by changes compared to the base level from year 1. Today, the four-eyes-principle is mandatory for a large portion of major and significant changes.

But that's not all. Thanks to improved training including simulated changes and by implementing a very strict six-month plan and many other measures, we were able to achieve a 79-percent reduction in incidents caused by changes. And this journey will continue.

We combat sources of error by learning from individual problems and using this knowledge to devise and implement programs for the whole organization. As we have said, the Quality Roadmap and systematically combatting sources of error are the central approaches in Zero Outage.

8.1   Quality Journey: Sending Big Initiatives on Their Way

Large, central initiatives can be planned and implemented on the basis of the Quality Roadmap. We have already mentioned an example relating to improvements in change management. But it is important not only to standardize and optimize on the process level, but also to implement improvements consistently on the behavioral and technical levels as well.

In this section, we will introduce the typical candidates for central initiatives. It is essential for any initiatives that are going to be implemented to be derived directly from the Quality Roadmap and thus the problems of the past.

First things first: If customer escalations are happening every day, the initial focus of Zero Outage must be on extremely fast and professional rapid action – namely, incident management. We have already explained the details of Zero Outage incident management. It ultimately comes down to quickly identifying critical outages, taking immediate action by following multiple approaches in parallel and, finally, guaranteeing good communication with the customer and the management teams of partners and suppliers.

Professional problem management is the basis for all **preventive measures**. This is not so much about an optimal process as it is about the right attitude amongst everyone involved. Mistakes must be addressed openly and honestly in order to derive the right preventive measures from them.

On this basis, the **main initiatives** can be defined each year according to the three P's (people, processes, platforms):
1. **Platforms/technology:** there are often weaknesses in patch and release management, for example when outdated network hardware has to be replaced, or no systematic failover tests are conducted even though they could provide a good deal of security, especially in a WAN environment. However, the most important component is standardization on every technical level, because diversity is the enemy of sustainable, affordable quality.
2. **Processes:** in nearly every company, there are gaps in configuration management when it comes to documentation – technical as well as customer-specific. The customer's "critical landscape" is not regularly updated.
3. **People:** the decisive factor for achieving Zero Outage is the attitude and mindset of the employees and managers on all levels. In addition to training, it

is extremely important to foster active communication and to sensitize senior managers so they set an authentic example of the desired behavioral patterns themselves. Concrete examples of this can be found in Chapter 15 of this book.

These were just a few examples of a **"quality journey"** over the course of a year. The actual challenge here is not to take on too much at once. If a program is too extensive, it can easily overtax the organization. This will result either in active resistance, usually with the justification that not enough human or financial resources are available, or in passive resistance, whereby activities are given a "green" status – meaning "completed" – when actually they have only been superficially implemented, with a corresponding limited effect.

On the other hand, the quality journey should be ambitious enough to have a tangible impact. If this is not the case, the change process for turning a company into a Zero Outage organization will not be noticed and will instead get lost among many other initiatives and changes.

When in doubt, the Zero Outage quality journey should be more ambitious rather than less ambitious. Resistance should be dealt with in a top-down fashion (from the management to the employees), and the scope should only be reduced if necessary. After all, what needs to be done should have been done long ago in accordance with ITIL. Getting rid of inherited liabilities also pays off by making incident management less reactive and reducing the number of escalations.

8.2   Proactive Fire Drills: Testing the End-to-End Chain

One important factor for the success of Zero Outage in a company is the speed at which operational disruptions are resolved. Every individual and organization learns through training and continual practice. During an emergency, the chain of alerts and the problem-solving process have to work reliably, quickly and securely. It therefore makes sense to **repeatedly practice for an emergency** to ensure there is no loss of routine or experience.

This is why emergency procedures are regularly simulated in the aviation industry. For example, pilots practice landing with an engine failure by letting the engine idle prior to landing and then trying to safely set the plane on the runway without gas. With each of these exercises, the pilot gains experience and learns how to correctly estimate altitude, find a suitable landing place and safely reach it in an emergency.

In a similar way, in the Zero Outage program incidents are simulated and unannounced alarms are triggered in various parts of a company. Measurements show how long it takes for everyone (including hardware and software suppliers) to get involved in troubleshooting, whether every management team responds promptly and appropriately, and whether the communication with all stakeholders is flawless. Every simulated disruption ("fire drill") ends by asking the participants about various aspects of incident management and holding a "lessons learned" session. Any shortcomings are openly communicated and logged to guarantee that these weaknesses are remedied by the responsible line managers.

8.3 Intensive Care: 360-Degree Support

For top customers who experience quality problems that can't be solved through the normal improvement measures for an account, and who need a clearly visible improvement in the short term, the 360-degree approach for a period of three months is a tried and tested tool.

In this case, several experts from the ICT provider are temporarily assigned to work intensively on the customer's most pressing problems – **outside of the regular line organization**.

This intensive care starts with a brief risk assessment. During this review, the typical sources of errors relating to processes, technology and employees are analyzed. The customer then defines his quality standard and which events are especially critical for him. Appropriate measures are then defined and implemented in a weekly cycle for the three P's (people, processes and platforms/technology). This could mean employee qualification or the acquisition of new resources, the improvement of core processes – especially in incident and change management – or the elimination of technical risks, such as exchanging outdated hardware. The success of these measures is checked regularly using customer surveys.

360-degree reports are a proven means of ensuring long-term operational quality for the most important customers. These reports cover all significant aspects of the customer relationship: operational quality as a basis, project quality (time, budget, quality KPI), service quality in order management and customer support, and the profitability and qualitative perception of the IT service organization in the customer's eyes.

Once again, **senior management** must take an active interest in these reports and put them to use. Employees must know that the management cares about the substance of the reports and values the work being done – this is the only way to guarantee that they will actively work on improving quality.

What does an intensive care "special forces" team look like? Here, too, it is important to take every level of the customer relationship into account. Various experts are therefore needed who can quickly and confidently identify and eliminate weaknesses if necessary: process specialists who closely examine supplier and service processes. Architects for the different technical aspects of a customer solution who can verify whether the solution's design and implementation still represent state of the art. And, finally, a supervisor – or customer head, as we say – who examines the organization for weaknesses on the customer and supplier sides, develops an improvement plan based on all identified problems, finalizes it with the customer, and then takes the lead in implementing it.

As already indicated, the **reason for dissatisfaction** often lies not with the supplier alone – **the customer side can contribute to it as well**. The customer head must therefore very quickly establish a trusting relationship with the customer's senior management to ensure that all problems are addressed openly and quickly, and that the necessary measures are initiated jointly in coordination with the supplier and the customer.

The central element of these improvement measures is the **service improvement plan (SIP)**, which answers the following important questions:

1. What must be improved? What are the precise weaknesses that have been jointly identified, and how have they been measured (result KPI)?
2. Which measures will be implemented to eliminate these weaknesses? What does the customer have to provide for this, and how will the progress of these measures be tracked (progress KPI)?
3. What are the target values for the result KPI that will bring an end to the SIP, and when are these results expected to be achieved?

Follow-up support is also important. After all, the SIP was necessary for a reason, and if this reason has not been permanently addressed, problems will arise again later. And if it comes to a third SIP, even the most obliging customer will be frustrated and wonder why escalations continue to occur.

Quality in Projects: Achieving Success through Standards and Transparency

9

Quality in projects is also critically important. To apply the Zero Outage principle to projects, you first have to know what the typical "pain points" are in a project, how to identify them and which measures can avoid or resolve them. The key to success lies in introducing **project standards for processes and products**, ensuring their implementation through proactive and reactive quality measures, and continually measuring the project's progress and results. In this chapter, we describe how this can be achieved in a Zero Outage organization. We conclude by condensing the most important rules into the ten commandments for project management.

9.1 Identifying Pain Points

The problem is inherent in the thing itself: a project is a unique, time-limited undertaking with a defined goal, and no two projects are alike. Every project's content, scope, approach and resources are different. Hardly a day goes by without another prominent, large-scale project becoming a talking point by failing in one or more aspects. The "classic" problems include:

- non-compliance with contractually agreed and previously communicated deadlines – we are not talking about days, but rather months or years,
- significantly over-running the originally calculated budgetary needs – not by the usual buffer of 10 to 20 percent for additional requirements, but by 50 to 100 percent or more,
- loss or replacement of suppliers and contractors during critical project phases,
- significant functional and technical deficiencies in the solution provided or
- inability to achieve the original project goal because important requirements were overlooked or incorrectly estimated.

In many cases, all of these problems occur simultaneously. New or misunderstood requirements cause deadlines to be pushed back, more resources are needed, costs explode, and so on. The reason for this is usually a combination of bad planning, changing requirements and under-estimated complexity.

IT projects, and software projects in particular, are unfortunately no exception here – in the context of such projects it is much more likely that requirements will change over time, a new IT platform or software component will trigger unplanned follow-up effects or result in more complexity, or that a particular type of solution never has been developed before. The uniqueness of the solution, often coupled with an initially incomplete picture of the target state (IT is expected to be flexible, after all), is the reason why a very high proportion of software projects will fail or at least be at risk. When a house is built, no one would think of moving in before the doors and windows have been installed. It would also be impossible to add a basement afterwards. But it is common practice in software projects to shake the very foundations of the software architecture whilst the project is underway.

Whether the client or contractor contributed to the situation, or what exactly was specified in the contract, are of secondary importance. Ultimately, this only influences the financial consequences for the parties. A failed project is still a failed project. Evaluation of a project's success, and thus the customer's perception of the contractor, is not shaped primarily by whether the contractual conditions were precisely met; the decisive factor is whether a solution that meets the customer's expectations is deployed in the end.

The **key to project success** lies in the following factors, above all:

- The uniqueness of the project is countered by the standardization of the project solution and project approach; there are suitable models for this regardless of the framework conditions.
- The project's requirements and framework conditions are comprehensively analyzed and described at the start, and there are clear agreements regarding how the inevitable changes will be handled during the project.
- Everything is documented in a project plan that is coordinated, communicated, consistently implemented and modified if necessary; outstanding requirements and the desire for flexibility can also be planned.
- There is a comprehensive and well-coordinated set of quality assurance measures – both in the project and from a neutral, external party.
- All aspects of the project's progress are measurable and continually reviewed; this transparency reveals potential non-compliances early on so countermeasures can be implemented.
- And last but not least, the know-how and experience of the project team are critical to success – particularly in key functions such as project manager, architect, development leader and test manager.

Taken together, these factors determine the quality of a project and the resulting solution. When correctly implemented, they generate customer satisfaction. But when unfulfilled or seriously neglected, they will cause a project to fail.

9.2   Use Standard Processes and Process Models

Although IT projects have a much younger history than projects in other fields, the people responsible for them do not start from scratch. Instead, they follow clearly defined and tested processes and standards for project management and software engineering. These describe which results should be produced by which project roles in which phases and stages of work. Global standards and process frameworks, such as the PMBOK Guide (project management body of knowledge) of the PMI, or CMMI (capability maturity model integration), form the basis for platform- and company-specific standards that make it possible to carry out IT projects professionally and with assured quality (see also Chapter 3).

The PMBOK Guide describes the phases of a project: initiation, planning, execution, monitoring and controlling, and closing. It also thoroughly illustrates the central knowledge areas, such as the management of time, costs, human resources and quality.

CMMI lays the foundations for all steps along the software development value chain, including customer requirements, the solution concept, technical design, programming, integration and testing, reaching all the way to overarching disciplines such as configuration management, quality assurance and training. It also covers the integration of project management and software development and the organizational and management processes required for this.

These standards create the framework for everything that must be done to describe and achieve the project goal. However, they do not specify how precisely or in what sequence the work should be done in a project. This is defined by process models. The three best-known models describe how to work according to (1) the waterfall principle, with sequential, completed phases, (2) an incremental principle in which each step builds on the previous one, or (3) an **agile process** characterized by many small, self-contained iterations.

Each process model has its advantages and disadvantages – there is no "one size fits all." This makes it all the more important to choose the process model that fits best with the requirements and framework conditions. For example, it makes little sense to choose an agile approach if an upstream project phase defined all requirements in detail and generated a functional specification. And vice versa, a strict waterfall approach will fail if requirements only emerge completely during the course of the project and every change will trigger a new change management process. This typically results in endless discussions about the original contractual arrangements between the client and the contractor. Not only does this have a financial dimension, it can also throw any project into chaos.

This is why agile process models – such as Scrum – are being used more and more often today. With this method, a requirements framework is defined at the start, the requirements are prioritized, specified in detail and then implemented in identical, often four-week cycles known as sprints. Each sprint delivers a complete, testable partial solution. The advantage of this is obvious: requirements are balanced against each other, complexity is easier to manage, parts of the solution are visible to the customer quickly, and you soon know whether you are going astray. It would be wrong to think that this model largely works without clear requirements, in an informal, non-procedural way. In fact, the advantages of the method are gleaned through a very disciplined, clearly defined and closely coordinated approach between the customer and the service provider. If this is not feasible, then even an agile approach will fail – but you will probably realize it sooner.

These process standards and models must be used to develop **a standard adapted to the portfolio and products of the company**. This involves defining the tools and templates that will be used. Here, too, standardization is essential – with the suitable type and number of tools that have been pre-configured for the standard. This also has benefits for the human factor: the more templates and tools are specified, the more standardization and fewer sources of error there will be. Tool-based automation – such as the generation of static and recurrent elements along the value chain – will further increase the quality and efficiency of the process.

How does standardization affect the project approach? The individuality of each project is not limited by this. Instead, by using process and tool standards, you can avoid having to re-invent the wheel every time. Flexibility and creativity are only reined in if they hinder or damage processes. The focus of the project is on the application domain, the technology and the project specifics. The job at the start of the project is to adapt standard processes so that the individual task can be solved – the keywords here are customizing and tailoring. This individual approach results in various plans that are coordinated with one another. The most important are the project schedule, communication plan, component integration plan, quality assurance plan, test plan and configuration management plan. These are combined and linked to each other in a project management plan. This – in combination with the solution architecture – is the blueprint for carrying out the project.

9.2.1 Implement Requirements Consistently

The customer's requirements are the focus – ultimately, every project will stand or fall by their successful implementation. Below, we have summarized the most important pre-requisites for professional requirements management.

- The requirements must be clearly documented, ideally at the start of the project or at defined points prior to implementation (depending on the chosen process model).
- There are functional and non-functional requirements. The functional ones are obvious; customers always state them when they describe the solution they want. The non-functional ones are less tangible but no less important. They should be defined just as early in the project. They include the stability, performance, security and extensibility of the solution. These requirements must be addressed, described and agreed. They are anchored in the system architecture and will determine the system quality – because this is primarily where projects fail.
- Before we can start the detailed functional and technical specification and the implementation, it is necessary to check that the requirements specification is complete and feasible and has been approved by the customer.

- It must be possible to test the implementation of the requirements – this is important for acceptance later on. Because they are so critically important, the requirements must be validated with the customer at an early stage. After all, the customer's formal approval of the requirements will mean nothing if the solution still doesn't meet his expectations. Possibilities for validation include desk checks, mockups, early prototypes and choosing an agile approach.
- Handling of new or changed requirements during the course of the project must be agreed with the customer and documented. This change management process is binding for both parties. This is an integral element in agile projects.
- In order to cope with changes and, ultimately, ensure the consistent implementation of requirements, it must be possible to keep track of every requirement – from its definition, through the design, to implementation, testing and delivery. Some of the questions to be answered here include: which requirements were implemented in the current version of the system? How good is the test coverage? Which parts of the system must be retested when a certain requirement changes? We refer to this as "traceability" – an obvious but difficult-to-implement requirement.

Another word about **non-functional requirements:** they are especially important because they bring together the requirements of the solution as well as its subsequent operation. This is partially why they are so difficult. These aspects are usually handled at different points in time by different parties on the client's and contractor's side. A Zero Outage organization is the ideal choice for bringing together development aspects and solution operability (i.e., project and operational expertise) from end to end.

The proper way to handle requirements is anchored in the process standard and a good project approach. The challenge is that these specifications must be followed consistently. A requirement will not be implemented if the key points have not been defined and agreed upon in advance – even if the customer says it's urgent and the developer happens to have time. To state it plainly once again: requirements management, regardless of whether it follows the waterfall principle or an agile model, does not just take place at the start of a project, it carries through the entire project – from the customer's initial idea to the successful acceptance. This only works if there is close, coordinated interaction between the customer and the provider.

9.2.2 Ensure Quality and Manage Risk

Quality assurance distinguishes between process and product quality. The right combination of both is essential for achieving the desired and agreed target state at the end of a project. The key factors here are:
- Quality is never free. It is always in a natural conflict between the available time and budget. This has to be balanced out sensibly.

- Everything can be planned, even quality assurance. If something has not been planned, experience shows that it will not be implemented – especially if time or budget pressures gain the upper hand.
- Good quality assurance starts on the first day of the project, not when the first results come in or the first delivery is due to be made to the customer.

Quality criteria must therefore be described at the start, and the target state and measures necessary to achieve it should be agreed and drafted. These are specified in the quality assurance plan.

Then things can kick off. The first **quality assurance measure** begins when the project is set up. It ensures at the start that everyone will work in compliance with the process, the necessary resources will be available and everything is planned. Training and coaching for the project team are supplementary measures. These are prime examples of proactive quality assurance measures, which should be followed by regular proactive and reactive measures.

Quality assurance measures take place both within the project and externally under the responsibility of a neutral quality assurance organization which is made up of experienced project managers, quality managers and software development experts. Neutrality and project independence are essential for everyone involved. After all, the goal of conducting a test is to find mistakes. In the event of a conflict – e.g., when quality comes up against time or budget constraints – there must always be an escalation path.

9.2.3 Conduct Reviews and Tests

Reviews and tests are the key measures for ensuring product quality in a project. They are statistical quality assurance measures conducted at a document or code level. Tests are dynamic measures conducted on the executable software that has been developed.

Reviews uncover the first, often serious errors at the earliest possible point, before anything has been developed. Error costs – that is, the cost of identifying and correcting mistakes – follow a very simple rule: the earlier an error is found, the less it will cost to correct it. If errors arise in the requirements or functional specification, these can be corrected directly at the document level (or in the modeling tool). Depending on their source, errors that arise during the acceptance phase may require a run-through of all previous project phases, or at least the development, integration and test phases. If errors are discovered in the design, these can be very costly – and this often endangers the delivery deadline. Consequently, no document should be approved for the next project phase until it has passed a review. Incidentally, error costs do not rise linearly with time, but rather exponentially – so it is worth investing here.

The Individual Test Levels
Tests are carried out on different levels, which are briefly described here:

- The first tests in a project are called **unit tests**. The developer conducts these in the development environment. A test is developed along with each piece of software. No software is approved for integration without passing the unit tests.
- **Component tests** are carried out during the first integration. These are designed to be very development-oriented, but they are not conducted by the developers themselves – this is where the four-eyes principle comes in. The first functional tests come into play during this phase.
- The **system test** is conducted when the complete software is ready for delivery, prior to acceptance and release for production. All test cases are conducted with the agreed level of test coverage. The specification of the test cases takes place with the functional specification of the solution, long before the solution is developed. The reason for this is obvious: the work takes place on the same level of abstraction – errors that are found during the test specification can be resolved cost-efficiently. At the same time, both the test costs and the necessary resources and time can be determined and planned.
- **Load, performance and security tests** are typically carried out in addition to this – think back to the description of the non-functional requirements. The tests necessary for this are developed in the context of the architecture and technical design. By this point at the latest, work should be taking place in a test environment that is very similar to, or ideally identical to, the production environment. If this is not the case, there could be unpleasant surprises during live operations.
- Prior to the go-live, a **system integration test** (usually the customer's responsibility) is conducted to check whether the developed solution functions in conjunction with other solutions. Here, too, it is necessary to conduct advance tests during development to check the interfaces to external systems.

Software is not considered production-ready until it has successfully passed these levels. Development cycles and re-work require repeat testing. Test automation is important here – in terms of repetitions, speed and efficiency.

All of this shows that nothing can happen without sensible planning, a dedicated and experienced testing team that is involved as early as possible, and the different test environments that are needed. The development, testing and operational teams work hand in hand. Alongside these tests, there are numerous key indicators that bring transparency to the tests as well as to errors and their resolution. If you don't know, measure and consistently use these indicators, you are "flying blind." We will address this in more detail below.

9.3 The Early-Warning System: Quality Gates

External quality assurance measures must also start at the earliest possible point. We encountered the first ones at the beginning of the project: these are the quality gates that continually ensure the quality of the process and thus, indirectly, of the product.

Quality gates in a project are **review and test measures that take place at the end of a development phase or level**. Failing at this point means that the "gate" cannot be passed and the next stage cannot start.

External quality gates can take place at any time. They guarantee that what was promised to the customer is implemented, in keeping with the Zero Outage principle.

Quality gates review in great detail whether the project results are achieved in compliance with the project management plan and in conjunction with the right project-internal quality assurance measures. These reviews are based on checklists geared towards the standard process. The reviews are conducted systematically. However, the quality of these checks depends on the experience of the quality manager carrying them out. Only a well-versed expert will notice if the work is performed not only in a formally correct way, but also professionally. Spot checks at the right points can reveal whether a result is at the necessary level of maturity. In cases of doubt, a review or comprehensive audit is carried out.

Quality gates result in the **identification of non-compliances and an agreement on measures and recommendations**. Everything that goes beyond a recommendation is pursued until a solution is found; until then, the quality traffic light stays red. The frequency and scope of non-compliances are an early-warning system for project crises. There are always signs before a project crisis, but you have to look carefully to see them.

In the context of Zero Outage, the **implemented quality gate concept revolves around four principles**:
1. Professionals with practical experience draw up the checklists. They make sure the right questions are asked.
2. Professionals with practical experience conduct the quality gate reviews. Only if there is mutual respect on both sides, can the weaknesses be identified and acceptance be achieved for the measures in the organization.
3. Neutrality, honesty and the dual-control principle: a consistent outsider's view brings attention to the problems; this is the only way that effective countermeasures can be implemented.
4. The questions on the quality gate checklists can only be answered with a definitive "yes" or "no." Any uncertainties are discussed in the context of the quality gates, and actions are initiated if necessary.

The quality gates are tailored to the service being provided. This means there are checks during the sales phase, quality gates for project management, software development, product creation, transition and operations.

Quality gates types depend upon when they are carried out:
- Initial gates are found at the start of a project and at the end of planning.

- Gates accompanying the project are found at regular intervals, e.g., every three to six months.
- Final gates before the end of a project ensure the transition to application management and operations; we refer to this as "service readiness."

Quality gates for projects primarily check the following:
- Was the project set up with the right resources and skills?
- Are the scope and requirements clearly formulated?
- Are change requests being managed correctly?
- Are the project plan and results up-to-date, and are both monitored regularly?
- Is there an adequate quality assurance plan in place, and are the planned measures being implemented effectively?
- Are deadlines and budgets being met?
- Are the customer's obligations being managed?
- Are the subcontractors adequately integrated and are they being steered accordingly?
- Are the communication structures clearly defined and are they being implemented?
- Are the key risks known, are there countermeasures in place for minimizing them, and are these being checked regularly?

Quality gates for products and platforms additionally address the following:
- Can the requirements be developed using the standard portfolio (products or platforms) in the requested time and at the calculated costs?
- Are the requirements of the market being planned in releases?
- Are all relevant stakeholders sufficiently integrated (development departments, partners, suppliers, marketing, etc.)?
- Is the product architecture sustainable and robust?
- Have widespread coverage and multi-client capability been taken into account in the structure and testing?
- Does the sales department have the right information for customers?

Quality gates reveal faults. The typical findings are:
- The scope was not adequately fixed with the customer.
- Responsibilities are not clearly defined.
- There is no adequate plan for achieving the project goal.
- Tests have not been adequately planned (measures, time and resources).
- The project team is thinking and acting too technically; there is not enough focus on the business requirements and effects for the customer.
- The partners/suppliers have been brought in too late and did not make it to do their job.
- The project is aiming at a 120-percent solution instead of an 80-percent solution. But the time to market is usually critical to success, and less is often more.

- Things are being complicated unnecessarily, important issues are not being distinguished sufficiently from unimportant ones ("nice to have").
- The team isn't saying "no" at the right points.
- The team stops at the borders of the organization, following the motto of "It's not my responsibility" (no end-to-end view).

The performance of the quality gates is measured using the **following key performance indicators (KPIs)**:
- Do the planned quality gates have the necessary scope?
- Are the checks being conducted as planned?
- Have the quality gates been passed/not passed?
- Are the agreed measures being implemented in the agreed time?

Ideally, the quality team will be centrally responsible for conducting the quality gate reviews. This guarantees neutrality and ensures that experiences and best practices are made available to everyone. Everyone can learn from everyone else.

It is the responsibility of the project and the delivery management team to ensure that quality gates are passed or, if not, that the necessary measures are implemented in due time. Responsibility for operational implementation cannot be delegated to the quality team. This is the only way to achieve the project goals set in the context of Zero Outage.

9.4   The Pièce de Résistance: Proactive Risk Management

Well-functioning, consistent risk management is probably the most challenging discipline in project management, and also the most important proactive quality assurance measure. The categorization of each undertaking starts with an overall risk assessment. This calibrates all of the tools for quality assurance, the necessary management level and reporting.

Project risks are known at the outset, and new ones will arise in the course of the project. Day-to-day project work includes tracking the project's progress as well as reviewing and assessing the project risks. The **probability of occurrence and amount of damage** are known for each risk. Risks with high damage potential and a high probability of occurrence are given special attention because they require countermeasures. Project managers should base their plans on a realistic, not optimistic, view:
- Risks with a probability of close to 100 percent are no longer risks but rather known problems that must be dealt with and are thus an integral part of the project plan.
- A risk of more than 50 percent means that it is likely that the problem will arise, so sensible countermeasures should be implemented.
- Financial evaluations must be taken into account in the expected project costs and weighted depending on the probability of occurrence. If this does

not happen, it can lead to "surprising" additional costs (overrun) when a risk becomes a problem.

Project risks must be clearly distinguished from plannable results. Every test will uncover errors – this is almost certain. For this reason, no project manager would only plan a test phase after the first errors have arisen. In keeping with proactive project and risk management, tests must be planned from the start.

The same applies to the fact that personnel shortages will occur in time-critical project phases if no back-up arrangements have been made, or performance bottlenecks or stability problems will arise during the first comprehensive system test. For these cases, too, measures can be planned in advance. Risk management deals with issues that should be avoided or, ideally, never arise in the first place. Good project managers will allow themselves **a certain degree of "paranoia"** and will continually deal with things that will hopefully never happen. In other words, a risk plan is not created at the start of a project and then shoved in a drawer; instead, it will be on the agenda regularly at project status meetings. Incidentally, if you never take any risks, you won't reap any rewards (see DeMarco et al. 2003). So, a certain affinity for risk is a fundamental characteristic of a good project manager.

Substantial changes in risks are reflected in the project's risk traffic light and are an important early-warning system for project crises.

9.5   De-Escalation Management for Projects

The measures described so far lay the foundations for a successful project through good planning, a solid approach and appropriate controlling. But what happens if something goes wrong anyway? We have seen enough examples of unforeseen problems and potential failure. An organization focusing solely on predictability and success is denying the reality of a project. Every quality organization needs an **emergency program and a rapid reaction force for project escalation** in order to avoid and combat project risks – what we refer to as de-escalation management. The route towards achieving this is also governed by planning and controlling.

De-escalation management is the counterpart to the early warning provided by quality gates. While these help prevent a fire ("smoke alarms"), de-escalation management is called in to extinguish a fire. Very experienced and assertive employees are critical to the success of both tasks. In a Zero Outage organization, they will form a single team or neighboring teams. De-escalation management has **all of the necessary competencies, authorizations and support on every level of management** all the way to the management board.

Here is an example of a project crisis we have seen many times before: The fixed-price contract for a development project has been signed, the key aspects of the scope have been defined. It is the customer's responsibility to provide the product platform, while the service provider implements the new project requirements, with the prospect of future maintenance and operations. The functional requirements have been specified, and both the project approach and milestones have been defined – in multiple development steps. Detailed questions about the exact services to be provided and the customer's obligations have not been sufficiently documented – but the parties agree on the project goal and have arranged for regular meetings to discuss the project's progress and joint steering. This works well for a long time – until unplanned situations arise. For example, certain requirements can only be fully implemented with a new version of the product platform. This initially looks like an implementation error which shows up in the customer test. It is clear that the customer is responsible for a new version of the product platform. But what about the costs of error analysis, re-development, new tests and the stability and performance effects that appear during migration? In any case, the discussion will drag on: what is the cause, who should have noticed it at which point in the project, who will pay for which costs, and who should be held responsible for postponements? Neither party planned for this. In hindsight, it is clear that they should have specified the product versions more precisely, planned for feasibility analyses, defined exact key figures for measuring performance, and much more – in particular, who bears which costs when unexpected changes arise. And the whole situation is much worse when the scenario is coupled with deficiencies in implementation and the non-fulfillment of the customer's obligations. Normal project work and pragmatic problem-solving are almost impossible in a conflict situation such as this. The project is in a permanent state of escalation, with claims and rebuttals. The shared project goal takes a back seat to contractual agreements and their interpretation. The deficiencies are obvious: on the one hand, a clearer agreement should have been reached in advance concerning services, customer obligations and non-compliances. On the other hand, effective (joint) crisis management is needed to solve the problem.

How can you identify a looming project crisis? Typical **indicators for project issues** include:

- The project reporting shows that defined threshold values for key figures have been surpassed – with respect to the budget, for example.
- There are subtle hints from the project team, such as, "I have a bad feeling…"
- Services are delivered on time, but the quality isn't up to standard.
- The customer complains that his expectations are not being met.
- There is a very high risk potential in the project right from the start.

In these cases, a detailed analysis of the project situation is carried out first to identify weaknesses, problems and causes. All aspects are taken into account (project team, processes, platforms/architecture/technology, governance). Based on the findings, the next step is to draw up a "back to green" plan which covers concrete measures for critical non-compliances and risks (who – what – by when). This plan is then monitored very closely (weekly or daily) by the de-escalation management team. Depending on the criticality, a weekly report or daily status mail may be sent to and discussed by a defined group of recipients.

Isn't this extra attention and additional reporting yet another burden for the already busy project team?

In practice, this **daily report** has been shown to significantly improve performance and lead to faster problem-solving. It ensures that, every day, every project member knows how far the agreed measures have progressed, whether they are working, where the critical path currently is and what should be done next. Everyone is confronted with the daily obligation to work effectively and efficiently on solving the problem and to immediately report any new risks. It is extremely important for the project reports to be

1. short and succinct; they should only cover the points that are currently relevant.
2. simple and comprehensible. If you can't clearly express yourself, you probably do not understand the issue.
3. reviewed and openly commented on by management. This is the only way to guarantee that everyone involved (project members and senior management) are suitably engaged with the project status and necessary measures.

Then adjustments can be made as quickly as possible. The central de-escalation management team provides the necessary support to the project team and ensures **direct channels for decisions, resources and prioritization**. And last but not least: effective crisis management creates trust – even in difficult situations. Fig. 9.1 illustrates a classic de-escalation process.

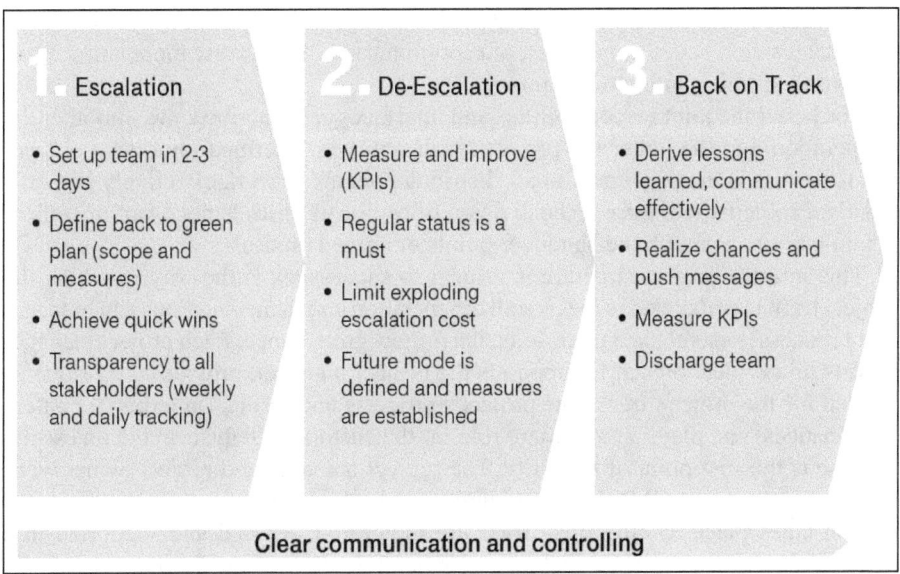

Fig. 9.1   Standard Procedure of De-escalation

Once the project returns to a sustainable normal state, the de-escalation management team hands it back to the regular project management team.

The reality is that the sooner a project enters an "escalation state," the greater the chances of quickly moving back into clear waters. The de-escalation management team supports the project – it is not a threat. A healthy error culture and consistent implementation are necessary.

9.6   Maturity Models: Project Organization and Governance

Part of the right project approach lies in choosing the **right project organization.** A project is traditionally divided into three teams: concept & design, development, and testing. Depending on the approach, these teams may work sequentially, overlapping or in parallel. Each team has one or more responsible individuals (subproject managers) who report to the overall project manager. The key positions on the team are the requirements manager, head architect, development manager and test manager. Quality managers, configuration managers and the project management office are responsible for cross-sectional tasks. If developments are divided chronologically into overlapping delivery cycles (releases), then separate release teams are formed, each of which is led by a release manager. The complexity of a solution can be functionally or technically reduced by setting up different development teams. This also enables development in distributed teams – especially **offshore or nearshore teams**. In agile development teams, responsibilities and functions are integrated and coordinated with one another as much as possible. An ideal team size is seven employees, plus or minus two. If there are more employees, it is worth splitting into more teams.

The better the joint responsibilities and interfaces are described, the smoother the cooperation and the better the project result will be. A defined joint development infrastructure is the pre-requisite for distributed teams to work effectively and efficiently. Problems will arise if the division of responsibilities is not clear, or if there is non-compliance with the handover points or agreed process.

This internal project structure is critical to success, as is the way in which the project team is embedded in the overall organization and in the functions and management processes established there – i.e., the project governance. Each project manager reports to a project owner. He appoints the project manager, provides the resources needed for the project, tracks the project's progress and results (in terms of content and finances) and plays an important role for the customer. In the event of an escalation, he is the first point of authority. The project manager and project owner work closely together. The CMMI standard defines the rules according to which this cooperation takes place. Both project roles are monitored, advised and supported in a successful Zero Outage organization by the neutral quality organization.

9.7   KPIs in Projects

The key performance indicators (KPIs) for controlling projects can be divided into three main categories:
- Project quality
- Project efficiency
- Customer satisfaction

For each category, there are different goals and resulting KPIs. Overall project success can only be achieved if the key indicators in these categories are balanced. If all project KPIs have been optimized but the customer isn't satisfied, then the project can't be a success. Equally, successful long-term cooperation will only occur if economic success is also assured.

9.7.1   Project Quality

In a nutshell, everything can be reduced to a formula that is communicated to the project manager at the start: time – budget – quality (the TBQ KPI). These three dimensions must be managed. Like other KPIs, they are related to and sometimes in conflict with one another.

Time
At the start, the project manager draws up a project schedule that is as realistic as possible, which is reviewed regularly and adjusted when needed. The primary concern is to reach the milestones in this schedule, most of which are contractually fixed with the customer. These are typically the completion dates for important project phases or delivery dates, usually the points at which a (provisional) quality-assured result is handed over to the customer. The "milestone compliance" time KPI measures compliance with agreed project milestones. It is measured for each project and across all projects – evaluated according to portfolios, customers, delivery units. Time is the factor that usually comes before all others, because a delay in contractually relevant milestones often incurs penalties and reduces customer satisfaction. Every missed milestone can affect an entire chain of dependent deadlines on the customer's side.

Budget
The project costs are calculated based on the project plan, usually monthly. The calculation is simple:

Total costs = actual costs + expected remaining costs

The actual costs are calculated directly from the finance systems used for HR costs, orders and infrastructure. The real challenge for the project manager is to determine the remaining expenditure and residual costs. For this, the project manager must systematically plan the project resources based on the work packages, estimate the expected remaining costs for their completion and assess the project risks. The simplistic mechanism generally used at the start is:

Remainder = planned – actual

This confirms the planned expenditure, and it must always be replaced with a realistic estimate of the expected remaining costs, including a content assessment.

The "budget compliance" KPI describes the (percentage) deviation between the calculated project costs and the planned costs.

Overrun = planned costs – calculated costs

Budget compliance and overrun can be represented in the project itself and also as highly aggregated data.

Everyone is familiar with the widespread **"90-percent-done syndrome"** in the critical end phases of projects. Most work packages that are not yet completed are evaluated as being "nearly done," and shortly before the deadline, work package after work package falls like a set of dominos. Time and budget can be professionally evaluated and monitored using what is known as the **earned value method**: each work package is rated according to its professionally-determined degree of completion. If the work packages are small enough, the rating will be only 0 or 1 – not done or done. This information is used to derive a "schedule performance index" (schedule compliance) from a scheduling point of view, and a "cost performance index" (cost plan compliance) from a budgetary point of view. An index of 1 means that the project is right on target. Values around 1 are accepted as normal fluctuations. Larger deviations above defined threshold values require corrective action. The advantage of this method is that it can be used from the start of the project instead of kicking in with the first milestones or shortly before the project's end.

Quality
Quality KPIs can be differentiated much like the measures for process and product quality.

The **process quality KPI** for "quality gates completed" indicates the compliance status for each project in the form of a traffic light. Green means that there are no non-compliances or all of them have been resolved. If this is projected and aggregated for all projects, it results in the process quality KPI. Zero Outage assumes that proactive quality assurance will always find non-compliances to resolve. The target value for the KPI is chosen accordingly.

Product quality KPIs are usually geared towards conducting reviews and tests and handling any errors found. There is a whole series of sensible KPIs that consistently build on one another and show the progress of tests and error correction in run-up curves. They measure:
- the tests conducted (absolute/percentage) and test coverage;
- positive/negative test cases and retests;

- errors categorized by project phase and severity;
- resolved and retested errors.

The KPI derived from this, known as **defect removal efficiency,** reveals the cost of errors. It depicts the relationship between the errors found (and corrected) in the project and the sum of all errors, including those relating to customer acceptance and live operations. A defect removal efficiency of between 85 and 90 percent is difficult to achieve, but worth striving for – from the point of view of both quality and efficiency.

Other quality KPIs review and evaluate the software itself according to pre-defined metrics. These check compliance with agreed coding rules, coverage through (automated) unit tests, program comments and software complexity, for example. These metrics can be measured in a largely automated way during the build process. With each build, the project manager and development owner for a module receive their current software quality KPIs; when implemented correctly, this happens on a daily basis. These figures are used during the development process for continual improvement. Such figures are helpful for the customer, too, for estimating and establishing the quality of the developed software.

9.7.2 Project Efficiency

Project efficiency does not mean carrying out a project within the agreed budget. The budget is what is calculated and agreed at the start. Financial figures such as the project margin, hourly rates and capacity utilization determine the economic success of a project and the delivery unit. But this says nothing about the efficiency that has been calculated into a project or how efficiently the work will ultimately be carried out.

For this, key indicators are needed that say something about **productivity in the provision of the service,** i.e., the effort to provide a defined service. Effort is measured in hours, person months or FTE (short for "full-time equivalents"). This is somewhat more difficult when it comes to defining the scope of service. Early approaches to this were based on lines of code. The advantage is that this can be easily calculated; the disadvantage is that the focus is on the programming phase; standardization is only possible within a programming language. The same program will require fewer lines of code in higher-level programming languages (such as Java) than in lower-level languages (such as Assembler). More recent approaches for calculating the "functional size" of a program use what are known as function points or counting procedures based upon them. These can be determined very early on based on the functional specification, regardless of the solution implementation. The key indicator reveals the productivity, measured in function points per effort, e.g., #Function Points/1,000 h or #Function Points/FTE. This is measured during the calculation (calculated productivity) and after the project is completed (realized productivity). When the project is implemented, productivity can be tracked (indirectly) through project progress indicators such as "earned value."

The productivity of a project depends on various factors, such as the development platform (the programming language comes into play again here), the use of standards, the degree of CMMI maturity, the stability of the requirements, the degree of automation, the team's experience, etc. Models such as CoCoMo (constructive cost model) define the framework for determining the expected productivity based on these factors. The organization must be sufficiently mature and experienced if reliable figures are to be produced. In the end, there will be a database of productivity values depending on the type of application, the solution approach and the implementation team. Increasing productivity means optimizing the relevant factors for success – and this can be measured continuously.

9.7.3 Customer Satisfaction

Last but not least, everything depends on customer satisfaction. No project can be successfully completed if the customer does not ultimately accept and successfully use the solution, even if the time, budget and quality demands have been met. Every customer will measure a project's success against the project quality indicators previously mentioned, or against other criteria, such as the attention paid to new requirements, the flexibility of the solution, proactive consultation from the service provider, innovation, cooperation with the specialist and IT departments, and much more. If these criteria are not known or are not regularly measured and discussed, the success of the project will be in danger from the start.

This problem can be solved by specifying these criteria with the customer at the start – usually not just for one project, but for all projects. There should also be an agreement on when and how these will be reviewed. An established and regularly updated **project or customer dashboard** ensures transparency across all agreed criteria – quantitatively and qualitatively. For quantitative measurement, all relevant project indicators are taken into account.

Qualitative criteria can be measured through **project and customer satisfaction surveys**. By regularly using and evaluating a questionnaire in a project, you can create a customer satisfaction index. The development of this in the dashboard can be tracked just like the time-budget-quality KPIs. Non-compliances can lead to measures for improvement that are discussed and initiated jointly with the customer. Problem points then become apparent in early phases of the project, and escalations and surprises at the end of a project are much rarer.

The same idea can be employed for several customers. A sensible number of project KPIs are recorded in a dashboard for all customer projects and regularly tracked – usually monthly. They show the development of project quality, efficiency and customer satisfaction in a Zero Outage organization.

9.8   Zero Outage: Ten Commandants for Project Management

Experience has shown that all modern methods of project management ultimately come down to a few ground rules that the organization must follow. This is the only way to achieve Zero Outage in projects. For example, here are the "commandments" that T-Systems has established as a reference for the entire company:

1. Our customer trusts our capabilities. Consistent and rigorous project management is the basis for achieving our project goals in terms of time, budget and quality and for maintaining our customer's trust. Our customer can rely on us.
2. Solid project planning is preceded by a reliable estimate of costs – following a standardized procedure. The costs of all services are estimated and documented using T-Systems' standardized method; they are the basis for project planning, determining the actual costs and estimating the remaining costs.
3. Project managers work with their teams to achieve business success – through personal and technical competence and by consistently implementing the project standards. All necessary roles and functions in the project are competently filled. All stakeholders are known and involved.
4. Project assignments and scope statements ensure clarity. Assignments are used to fill the key positions in a project, such as the project manager and quality manager. The scope of projects and sub-projects is specified in scope statements.
5. At every point, there is visibility about where the project stands – thanks to consistent planning and progress measurement. The project planning is well structured, completely traceable and always up to date.
6. Consistent requirements management ensures that customer requirements and changes are traceable until the deliverable.
7. Quality gates ensure success based on company standards. Every project adequately plans its quality gates, which check the approach and results at defined points in time and lead to corrective measures if necessary.
8. Risks are identified promptly and managed consistently. Known risks are evaluated; effective countermeasures are assigned to them and tracked regularly.
9. Subcontractors are viewed as a critical factor for overall success and are managed accordingly.
10. The customer's obligations are critical to the success of the project. Compliance with these obligations is planned with the customer at the start of the project and monitored regularly.

From Customer Perception to Customer Satisfaction 10

The previous four chapters dealt with key factors that can positively affect customer satisfaction when implemented thoroughly: an effective organizational structure, high-quality ICT operations and professional project management. This chapter will round this off by focusing specifically on customer perceptions.

In a globalized world, it is imperative (and therefore extremely common) for all goods and services that are traded between companies to be standardized, measured and comparable. Every ICT provider must therefore work with their customers to define meaningful key performance indicators (KPIs) that characterize the provider's services for the customer as accurately as possible. These KPIs then serve as helpful measurements for the quality of the services provided. But despite all of these efforts, one thing is clear: quality is in the eye of the beholder. In practice, many other **"soft" factors** are at least as important to successful cooperation as the KPIs which are specified in a contract – for example **a personal relationship with the customer**, transparency in solving unforeseen problems and the **quality of communication** on all level of hierarchy.

10.1   The Service Manager as the Interface to the Customer

It is not uncommon for customer satisfaction to decrease over a long period of time, even when service level agreements have largely been fulfilled. This is why it is critical for each customer to have a personal point of contact in the provider's company. This point of contact is the service manager, a strategic component of ITIL core processes, who is involved in all communication – both internally with the delivery units and externally with the customer. The service management team is responsible for the contractually agreed service performance, meaning that this is another **end-to-end responsibility**. The term "service management" represents the entirety of a company's professional ability to offer the customer added value through services – in other words, all functions and processes that are necessary to manage the life cycle of services. The goal here is to manage resources so that they generate the greatest possible benefit for the customer.

Large application organizations use ITIL as a guideline for structuring their service and support concept (see Chapter 3). But why should customers and providers comply with these standards? It's very simple: because they guarantee comparability and visibility. They make it possible to compare offers and choose a suitable provider. However, such standards provide relatively little help when it comes to establishing truly good service management. They are a pre-requisite, but merely "fulfilling" these standards is not enough for a really good customer relationship.

Functioning IT governance is the cornerstone of a coherent overall service management concept. This is the context in which goals are defined for measuring the configuration of a company's IT (hardware and software) and associated processes. Other functions include the organization, control and monitoring of the IT landscape and its interfaces.

The potential goals may be:

- increasing efficiency
- improving service quality
- boosting customer satisfaction
- making risks more transparent
- a periodic overall review of the IT strategy and operating processes

Successful service management is characterized by a strategic approach combined with a keen understanding of the current challenges facing each customer. To be successful, service management must also understand how important the provided solutions are for the customer and which process depth is required for controlling high-quality services. The service management team must be able to approach the customer and understand his priorities and actual needs. Proximity to the customer is critical. And the service management team bears the responsibility – internally and externally – at the interface to the customer.

What are the specific responsibilities of the service management unit, and how should this unit ideally be structured? In the following paragraphs, we will look at the responsibilities of a service manager using the "order-to-cash" process as an example:

Order Management

A customer's order is the result of an existing business relationship between the customer and the service management unit. Service management is specific to each customer, so the team understands the customer's processes – and, even more importantly, the requirements and benefits of the solution for the customer's business. The order is processed and implemented in accordance with the contract.

Implementation and Handover

After the customer has placed an order, service management is the internal contractor for the operating units and external partners. The service management team draws up the final specifications and uses quality gates to check the implementation until the handover to the customer.

SLA Management

Operations are usually controlled by a service level agreement (SLA). Internally, SLAs are secured by corresponding operational level agreements (OLAs). Service management monitors compliance with the OLAs and SLAs. At least once a month, a **service review meeting** is held with the customer to evaluate the fulfillment of the contractual service levels. The service management team is responsible to the customer – including when contractual SLAs have not been fulfilled. If SLAs have been violated, the service manager takes the lead in a service improvement program (SIP) and supervises this process for the customer.

Incident, Change and Problem Management

ITIL Foundation is the current industry standard for operational processes, and it should be established in the company. It is advisable for ITIL-compliant processes to be introduced on the customer's side as well to ensure the smooth running of procedures. The service manager must therefore be ITIL-certified to ideally supervise processes relating to incidents and changes.

Commercial Management

The service management team bears overall commercial responsibility for a customer. This means the service manager is a full-fledged point of contact for the customer with the corresponding decision-making authority.

Claim Management

Efficient and customer-oriented claim management is another challenge for the service management team. The service manager must bring a high degree of technical expertise to claim negotiations and discussions to ensure a viable customer relationship now and in the future.

Review and Continual Service Improvement

This covers the evaluation and optimization of IT services. Business processes can be evaluated with the help of previously defined key figures. The focus should be on key figures that represent the customer's critical success factors. This process is supported by various reviews relating to IT service, SLAs and processes.

IT services are put into operation for transition projects. Project acceptance means that a process has been successfully implemented. However, an active improvement process in the company is tremendously important and should lead to the continual development of existing processes to optimize customer benefit and ensure customer satisfaction.

10.2   Detecting Service Management Weaknesses

What are the most frequent service management problems – and how can you tackle them?

Problems are most commonly caused by **a lack of monitoring systems and corresponding KPIs as well as inadequate reviews of the continual improvement process**. Customers simply have no visibility into what is happening on the provider's side, which can quickly lead to mistrust, protracted discussions and, in extreme cases, escalations.

Other challenges may include **unreasonable service level management demands**. For example, 24-hour service might be feasible for one customer but not affordable by every customer. The service manager must compensate for this fundamental detail in a contractual customer relationship by working with the customer to prioritize the most important systems and company processes. The service manager should also give customers the feeling that they are always completely informed of any remaining risks. With this shared understanding, it might be possible to offer a focused 24-hour service, for example.

Maximum availability as well as incident and capacity management are additional challenges for service managers. For example, it may be necessary to determine the needs of a customer who does not have consistent business requirements but rather seasonal ones, and who requires support with monthly statements or seasonal sales. Ensuring that users on the customer side accept the work of the service provider is another responsibility that can only be tackled jointly. Employee qualifications and fluctuation are also frequently mentioned as causes of problems.

How Should You Deal with These Challenges?

Weaknesses such as this, and any necessary adjustments to IT services or service management, must be discussed with the customer and described from the customer's point of view. Competition-critical applications must be identified and named. Furthermore, the service level requirements should be described in detail. Employees should be qualified and certified in accordance with industry standards. It is essential for employees in service positions to receive training in typical industry standards or even ITIL certification – through an accredited inspection body or other well-known auditors. Furthermore, IT governance should be clearly coordinated and documented between the client and the contractor in order to avoid misunderstandings and to clearly regulate tasks, responsibilities and the customer's obligations. The hallmark of a well-structured account is a viable model for cooperation between the provider and the customer. **Service management is a "people business,"** meaning that it is an important part of customer/supplier governance. Ideally, the internal account organization should reflect the customer's organization. Additionally, these "rules of engagement" should be continually expanded and improved together with the customer.

Fig. 10.1 shows an **example of a global governance model**.

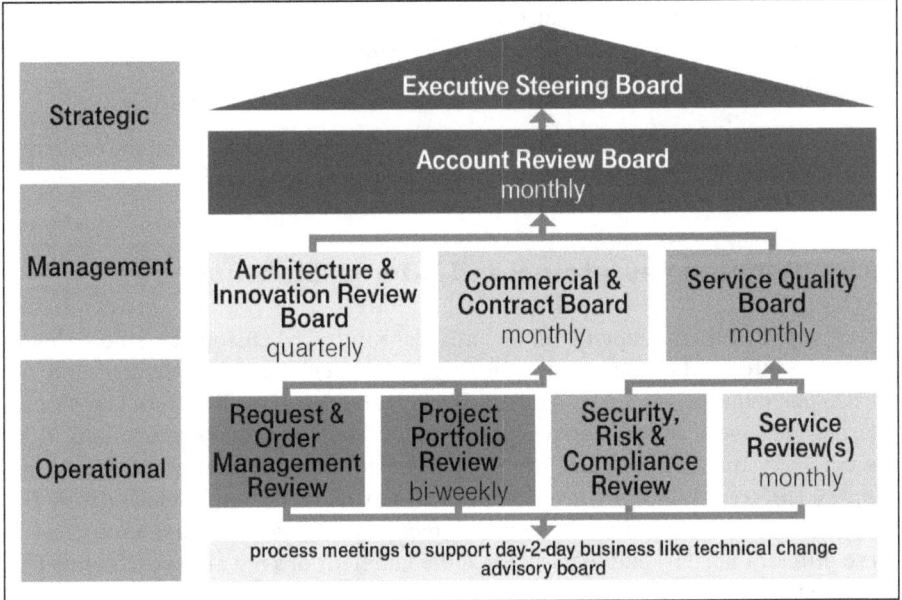

Fig. 10.1   Example of a Global Governance Model

Operational Level
Service management guarantees and actively supports communication about daily business. Monthly quality reviews are a permanent part of this cooperation, as is a forward-looking view of the services and the goals of the customer.

Management Level
The operational level prepares regular meetings with a set agenda to be held with the management team. The management level looks at the past and coming six months and serves as an escalation level. It also prepares the strategic governance plan.

Strategic Level
The strategic level is the highest escalation instance in a well-structured governance system. It is forward-looking and specifies the strategy for the coming six to twelve months. In addition to bearing responsibility for the above-mentioned processes and service levels, successful service management offers particular value by playing an active problem-solving role. The service management department thus takes on comprehensive, end-to-end responsibility.

What Does Service Management Look like in Practice?
Successful service management establishes a functioning interface to the customer and continually optimizes it. In our experience, the following steps are the best way of doing this:
1. Define the most important internal and external stakeholders

2. Active SLA management: question and modify the SLAs
3. Listen to customer feedback and handle it successfully
4. Set up a continual improvement program

These four steps should be carried out at least once or twice a year with the customer.

10.3 Defining Key Internal and External Stakeholders

Every responsible individual – on the provider's side and customer's side – should always be informed of the events affecting him or her. It is therefore advisable for the account managers to keep an internal stakeholder list. This list should include all internal employees and executives – from the different supplier organizations, from the company management and from departments such as finance and human resources. **Different hierarchy levels require and expect different levels of detail**. For this reason, regular reports and recurring meetings should be designed to cater to these different needs – and to ensure that the target groups are always integrated.

Similarly, there will be stakeholders at every level in the customer's organization who must also be included in these lists on a regular basis. If stakeholder management does not work perfectly on any of the levels, there will inevitably be problems in the customer relationship. This is true even if "only" a lower level is affected; problems will be escalated quickly, and the respective supervisors will receive "tainted" information. Furthermore, problems trickle down rapidly.

The two stakeholder lists can be used to develop a communication plan which keeps everyone permanently informed of changes and results.

10.4 Active SLA Management

Service quality is objectively measured with KPIs. Service quality standards are agreed upon in contractually defined SLAs. But in addition, there is always a perception of service quality on the side of the customer which can significantly influence customer satisfaction. This is not based on an objective, factual evaluation of the services provided, but rather on the experience of cooperating with the provider and using the services on a daily basis.

Various service levels have been established in the IT industry for objective measurement. We believe the four critical ones are:

Availability of Services
Regardless of the type of service that has been agreed, it is absolutely essential for the customer to be able to use the service. For example, if a bank offers an online portal, its customers must be able to access it. Availability is therefore fundamental to SLA management and part of daily service quality monitoring.

Recovery Times

Everything must operate flawlessly: hardware, software and IT processes. This is the top priority when it comes to quality requirements. From a technical point of view, extremely high availability can be achieved, even if 100-percent availability is not possible in IT. IT operations always run the risk of disruption – for example through hardware or software errors, viruses or cyberattacks. But the way such incidents are handled reveals a great deal about the quality of the provider's troubleshooting processes. Therefore, in the context of the service levels, it is crucial to define recovery times and, if necessary, to prioritize recovery operations in the event of an incident.

Performance

KPIs should be defined between the customer and the service provider to measure the performance of the services that are provided – this is a must. These KPIs may be based on application reaction times, the response times of individual processes, delivery times for hardware, or even consultancy services. One way or another, they are absolutely fundamental to cooperation. This is because service quality can be measured on the basis of these KPIs. KPIs that are measured repeatedly – daily, weekly and monthly – are ideal reference values and a basis for establishing programs to objectively increase or stabilize performance. KPIs create a high degree of transparency and offer neutral values for measuring the actual quality of services – this might provide arguments to address any negative perception of service quality by the customer.

Reaction Times

It is irritating for any customer to have to wait on hold for a long time on a telephone hotline. But reaction times for service desk/help desk environments can be clearly described, agreed and measured. By measuring something, we can analyze it, improve it, measure it again and continually review it. What's important here is the "first time resolution rate," or the proportion of tickets that can be resolved directly by the service desk as well as the proportion of tickets that cannot be resolved within the SLA. In the case of critical issues or mass disruptions, it is vital for the appropriate expert teams to be alerted immediately; in Chapter 7 we described how to systematically guarantee high reaction speeds and visibility for the customer.

10.5   Dealing Successfully with Customer Feedback

Customer feedback about service and quality is needed to plan and implement optimization measures. Only the customer can communicate his own perceptions, and this information can be used to develop an individualized improvement process. For example, while accuracy and traceability are the top priorities in pharmaceutical production, time-to-market is more relevant to investment banking. Through a regular, structured exchange about projects and important operational factors, improvements can be identified for each situation – and the KPIs can be modified accordingly.

In practice, customer feedback can be gathered well during **regular service discussions**, and it is important to ensure that the customer is continually informed of the implementation status of his requests. These discussions are also a very good opportunity to raise other issues or determine the upselling potential as a service provider. After all, every comment, suggestion and even complaint from the customer is a chance to improve cooperation.

A successful service discussion will therefore address current customer feedback and the status of any earlier feedback. These joint discussions should also be used to **clarify or correct any ambiguities in supply and service relationships**.

It is important to understand the customer's perspective in this context:

- Is this a new account for the service provider, meaning that trust-building activities must first be completed on both sides?
  - Particularly at the start of a contract, it is advisable to talk very regularly about supply and service relationships and to make corrections early on.
  - Ambiguities in the specification sheet or contract should be identified and mutually clarified or explained in detail.
- Has the customer recently experienced a disruption?
  - No customer will be able to pay much attention to a service manager if they are currently experiencing a disruption or if they recently experienced a disruption. In this case, the service manager should specifically address the disruption again and ideally present the initial findings of the root cause.
  - This opportunity should also be taken to clarify if the customer was satisfied with the communication during the disruption. After all, when it comes to troubleshooting, the most important aspect besides responsiveness is proactive communication so the customer always knows what is happening.
    - For example, should employees be sent home if the disruption is going to last for a long time?
    - What steps is the service provider taking to fix the fault?
    - Should the customer have his own technicians on hand to provide support during an emergency change or test the result of the change?
  - Has anything changed in the customer's organization or environment?
    - If points of contact have changed or different organizational units are now responsible for the contract, new relations will have to be established with the customer.
    - Has there been a company merger or fundamental change in the customer's strategy to which the provider must react?
    - Is the customer himself currently undergoing an audit that could be supported by the service manager?
  - Are contractual negotiations with the customer on the horizon?
    - The customer is looking for arguments for a price reduction and is strategically increasing the number of comments or complaints.
    - The customer is trying to expand the agreed scope of services.

So, feedback from a customer should be judged based on the situation. When we **understand the reasons and background conditions,** measures can be adjusted accordingly. It is also advisable to document any agreements made with the customer in this context so that regular, objective reports can be provided about the agreed measures.

10.6   Improving Continually[1]

The goal of quality management in an IT service organization must be to continually ensure that the customer is satisfied with the services he has received. To this end, it is essential to systematically, permanently and objectively measure the degree to which all of the customer's major needs have been fulfilled. Based on these findings, measures can be implemented promptly (if needed) and regular, professional communication of the results can be guaranteed.

Both the **delivered service quality** (objective service quality) and the customer's quality demands, which are reflected in his **subjective perception of service quality,** will continually change over time. A **continuous improvement program** – that is, the ongoing improvement and optimization of the service quality level – is therefore a key factor for a successful long-term customer relationship.

10.6.1 Typical Development of the Perception of Service Quality

Although service quality is measurable objectively, it **will change** over time. In part, this has to do with changing requirements. At the start of the system's life cycle there might be many adjustments and changes if, for example, not all of the customer's requirements were taken into account in the development and project phase. Additionally, requirements can change if the underlying business model has to be adjusted or if the technology is not yet as mature as it would be in later phases.

Employees working with the system will also gain more experience; documentation will improve, and weaknesses will be rectified. Operational processes will be optimized, and troubleshooting routines will be established thanks to this documentation. Deployment of a system is usually followed by a stabilization phase on the customer's side; fewer and fewer changes will be necessary because the underlying business processes will have proven themselves and become stable. All of these factors have a positive effect on the subjective perception of service quality.

These positive factors are countered by a number of negative effects. In terms of technology, hardware components will increasingly fail on all levels over the course of years. Processes will become more bureaucratic, leading to limited flexibility and long lead times for changes. There is a danger that, after long phases of stability, employees will stop paying close attention. This can result in carelessness in change

---

[1] This section is based on Kasulke 2014.

quality, patch and release management, problem management and customer communication. Additionally, in the last third of a system's life cycle, the customer will have new requirements that are difficult to meet with the existing technology.

Along with these objective factors, **the subjective expectations of the customer will change as well**. While the customer might have tolerated the odd outage or deficiency at the start of the life cycle, as the system grows more stable he will increasingly expect services that are nearly 100-percent reliable. Both the customer and the underlying business processes will therefore become less tolerant of outages. After five years, there is usually a considerable discrepancy between the original service level agreement and the system availability that is expected.

Guaranteed objective service quality should be the basis of any successful customer relationship. Additionally, the subjective perception of service quality is highly relevant to a long-term relationship between an IT service provider and a customer, and the customer will typically become more and more dissatisfied unless some action is taken. This is because he will increasingly take the contractually agreed services for granted, and any disruption caused by one of the above factors that has a negative effect will make him dissatisfied and lead to mistrust. In response, the IT service provider will launch a quality offensive – a service improvement program (SIP) – in order to fulfill the customer's expectations once again. The SIP addresses the negative influences – such as outdated hardware or errors in change planning, preparation and execution – and thus ensures an objective increase in quality to meet the customer's expectations.

In summary, the **subjectively perceived service quality often follows an up-and-down trajectory** – that is, a decline in the subjective perception of quality over time, followed by a SIP that temporarily improves the quality, followed again by a decline.

10.6.2 Early Indicators

The decline in both objective and subjectively perceived service quality can be identified early using six decisive indicators that are described in Tab. 10.1. These must be reviewed for each customer regularly – three to four times a year.

Tab. 10.1   Risk Factors, Effects and Checkpoints

#	Risk factors	Risk/effect/result	Checkpoints
1	Unexpected (high) profit in unchanged contract > 6 months in a row	With steady revenues, costs were probably cut (in the areas of materials/human resources). Reduced quality by the provider means customer is no longer engaged and will look for another provider.	Customer-related revenue planning (rolling forecast)/ 12 months
2	Changes to the customer's organization structure	Contract or parts of it could be understood/interpreted differently. → This leads to complaints and/or other demands.	Quarterly figures/ quarterly status of account
3	Contract or parts of it expire	1. Departure and search for new prospects/ employers among internal employees → The best go first. 2. Rejection/decline in awareness/behavior among internal employees and on the customer's side	Proactive measure prior to expiration of contract
4	Governance model is no longer followed (cancellation of regular meetings)	Decline in professional work → risk of incomplete work (projects/programs) → interpretation → "I've heard..." → personalization → escalation → relationships	Quarterly figures/ quarterly status of customer/account
5	The key players perform the same function in the same account for longer than three years	Friction losses → decline in driving force of innovation, architecture, growth and quality	Regular/annual evaluation and assessment of employees in their functions
6	Existing programs/measures are not sustained by the customer account managers	Declining development/less progress	Quarterly status of account

It is advisable to actively implement measures after 18 months at the most, even without these tangible reference points, to ensure long-term customer satisfaction.

10.6.3 Measures Aimed at Continually Improving Objective Service Quality

As stated, securing the objective and subjectively perceived service quality is an important goal for IT service providers. Various approaches and measures can be taken to guarantee the important foundation of objective quality – such as an optimization program. At the start of an active continuous improvement program, all existing or newly identified weaknesses relating to a customer are systematically tested. Fig. 10.2 shows the main areas to investigate following the classification system of processes, platform/technology, program and people.

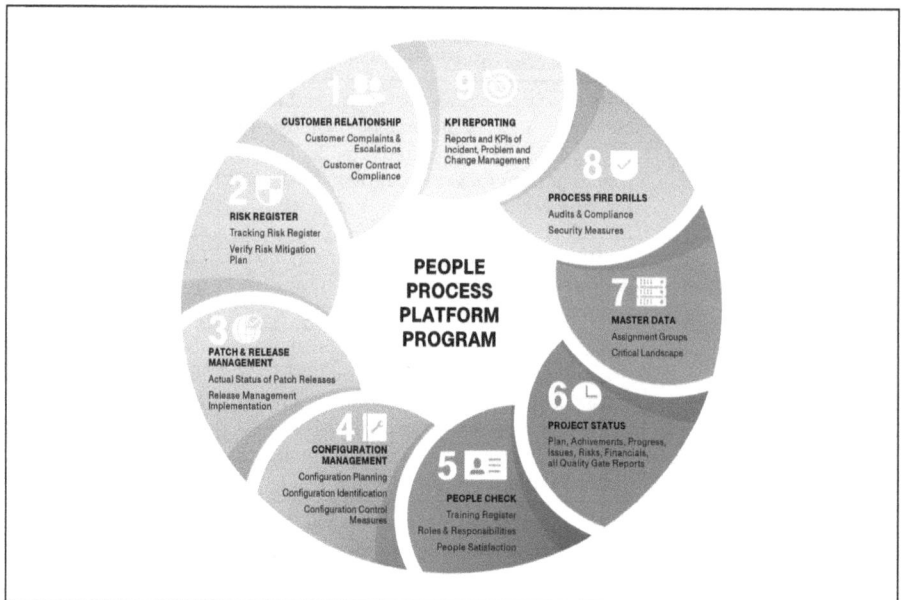

Fig. 10.2 Systematic Check of a Customer

10.6.3.1 Processes

Configuration management is a good starting point for improving processes: an accurately maintained configuration management database (CMDB) is a useful indicator for determining whether the processes in change management, patch and release management and monitoring are working and applied in a disciplined way.

The reporting of key performance indicators and – above all – documents for the regular service review board with the customer are a valuable reflection of how seriously governance is being taken by both sides. This can reveal the extent to which the customer's expectations are in line with or deviate from the results and the SLAs and how visible these results are to the customer.

A systematic test (**health check**) should take no longer than four weeks and include a sampling of all key elements of the customer relationship. The results of this test are translated into a quality improvement program that systematically rectifies all weaknesses in processes, people, and platforms/technology.

10.6.3.2 People

In IT in particular, the human factor is critical. Now that processes (CMMI, ITIL) and technologies have been standardized, the only way that IT providers can positively differentiate themselves from the competition is through their employees. IT consultancies have known this for a long time, but in IT operations the focus is often still on processes and technologies. Below, we present important measures that can help IT service providers **to ensure that human error is avoided and the level of alertness is kept high over time**.

Rotation Principle

Everyone experiences a learning curve in their job. When someone is given a new task, they first go through a familiarization phase. They have to understand the processes, their responsibilities, their colleagues and the technology, and they gradually acquire these skills. Depending on the complexity of the task, as well as the employee's prior knowledge and learning abilities, it will take six months to a year before a new employee reaches full productivity. The learning curve is initially steep, but gradually flattens out over time.

Employees grow more self-confident as their work becomes routine and the quality of the work rises. But this self-confidence often leads to increased carelessness, meaning that employees deviate from or neglect processes; experience may partially compensate for this, but the frequency of mistakes rises in absolute terms nonetheless.

In practice, the three-to-five-year rule has proven useful for ensuring that the company benefits from growing experience and routine while preventing familiarity with the task from resulting in carelessness. The three-to-five-year rule means that a manager or employee should usually not stay in the same position for the same customer for less than three or longer than five years.

Regular Appeals to Raise Awareness

A way of raising employees' awareness of quality and compliance with processes is to conduct well-planned and executed events, such as meetings of all employees and managers on a customer account or, – in the age of globalization – teleconferences or web conferences. In the best case, representatives of the customer will speak directly with the employees of the IT service provider in order to share information about the importance of the customer's main applications and the effect of disruptions or project delays. Given the extensive division of labor and the fact that multiple organizational units and companies are involved in the supply chain, it is very effective when a relationship has been established with the end consumer and every employee is aware of their specific contribution within the big picture.

10.6.3.3 Technology and Tools

Hardware and Software

Technical infrastructures age quickly. Hardware has a life cycle of three to seven years, but when it comes to software, a complete overhaul is usually required on every level within just one year; after this point, there is little or no support from the manufacturer, and security vulnerabilities could arise. But this is countered by the risk that very new software releases often harbor unknown faults, meaning that early adoption or frequent updates can lead to outages.

In the continuous improvement process, a strategy must be defined for each class of hardware and software assets to find the perfect balance between these two factors. These rules must then be implemented and monitored. An annual review of the chosen update intervals will guarantee that changes in the release policy of individual manufacturers are taken into account and the experiences of other companies using the same hardware or software can benefit the customer.

Disaster Recovery (DR)

In an age of multiple redundancies on all levels of hardware and software, customers should never experience an outage of their high-availability applications. In practice, however, redundancies sometimes fail to kick in, so a customer may experience serious operational disruptions, despite all of the technical precautions. Frequent reasons for this include:

- no consistently separated layout of connections,
- inconsistent software patch levels on redundantly designed clusters,
- single points of failure (SPOF) in the basic infrastructure (cooling including control, power supply and uninterrupted power supply [UPS]) in the application design or in the network structure.

Furthermore, the behavior of an entire system changes over time due to changes on each level (e.g., changes in usage, application software, network structure, hardware or system software).

Therefore, a continuous improvement plan should include regular reviews of a customer's critical landscape to check whether the redundancies are functioning for all of a customer's important systems. This happens in two steps:

- First, existing documentation should be used to carefully check a system on every level (basic infrastructure, hardware, network, software, application level) for any inconsistencies, known errors in the release and patch levels, or single points of failure. This should happen at least once a year and must be considered essential preparation for the second step.
- In the second step, the systems are checked in the context of changes to see whether the redundancy is working; in other words, major outages of sub-components are simulated at times agreed with the customer to test the system's behavior during an emergency. The results are then taken into account for further improvements. It is advisable to take a step-by-step approach here, meaning that only outages which can be reversed rapidly should be tested to begin with.

Alternatively, real outages of sub-components can be evaluated and recorded in the test plan as (involuntarily) conducted tests.

Ideally, there will then be an overall plan for the customer which notes the time and result of the latest availability test for each key system.

10.6.4 Measures Aimed at Continually Improving Subjective Service Quality

Quality is in the eye of the beholder. In addition to "hard" quality criteria such as operational stability and reliability in projects, it is essential to manage the subjective perception of quality on the customer's side. But how can you do this?

The customer's perception of quality originates at the interfaces between the IT service provider and the customer – that is, at the service desk, in the service review boards, during service requests and in joint project work – particularly among the steering committee and in the requirements and test phase.

It is essential to present a very well-prepared front at these interfaces and give the customer the feeling that all of his needs are understood.

This can only be achieved through an intensive, comprehensive training program that prepares every employee for the responsibilities they will face if they are in direct contact with customers. Key topics should include technical training, as well as behavioral training, i.e., how a service delivery manager should behave toward the customer in the event of a disruption, how claim management can be handled well, or how to optimally prepare and conduct a service review board.

The customer's satisfaction with the service desk is measured regularly as the basis for a targeted improvement in the resolution rate and customer satisfaction.

10.6.5 Success Monitoring and KPIs

To monitor success, it is advisable to define a dashboard with all of the result and progress KPIs that are important to the customer and the management – with starting and target values as well as monthly and weekly goals. The result KPI shows results that have been achieved, e.g., the number of disruptions in each class, the time to resolve problem management tickets, customer satisfaction with the service desk, and TBQ compliance, meaning the extent to which time, budget and quality demands were met in projects.

Progress KPIs measure whether the defined measures were implemented, such as how many fire drills were carried out.

The dashboard must be defined together with the customer and reviewed with him on a monthly basis to ensure that progress is totally transparent for the customer and the measures reflect the customer's main priorities.

The Challenge of Statutory Regulations

The growing extent of legal regulations affecting IT, telecommunications, security and data privacy means that compliance is becoming an increasingly important issue. And because legal compliance is the basis for Zero Outage, it is something that should always be discussed in detail with the customer. Failure to comply with statutory regulations can result in fines, loss of orders and loss of reputation. Unfortunately, regulations and laws are normally very industry-specific, and this can present a challenge for ICT providers. For example, pharmaceutical companies and banks are subject to the additional stringent regulations and documentation requirements of **GxP guidelines** or **ISAE 3402**, while companies in other industries only need to show that they have an IT provider who complies with the applicable ISO standards. Statutory regulations usually exist to make sure that work is being carried out properly at all times and that accurate records are being kept. Knowing this, the IT provider and his customer can then decide what measures to introduce, what standards to apply, what documentation and reports to produce, what audit cycle to implement, and what criteria to use to verify compliance with the legal requirements of that particular industry.

11.1 The Example of the Pharmaceutical Industry

The pharmaceutical industry has to account for every process step within its operations. For example, it must check compliance with hygiene regulations, accuracy of processing, dosage of its ingredients and also, of course, what staff were involved in the manufacturing of each batch of product. Any changes in production or in the equipment used must be accurately documented and confirmed, using the four-eyes principle. The operator of a pharmaceutical company's ICT systems is also required to show in detail how each of their processes operates. An IT provider is not a pharmaceutical company however, and not all parts of the IT system are necessarily involved in the production chain. For this reason, it is advisable to agree in advance exactly what requirements each IT system has to meet and what supporting documentation should be provided.

Let's use the **analogy of an onion to describe the collaboration of IT with production**. IT is in the outer layers, which are not directly related to the production of pharmaceuticals, with other departments such as financial systems, human resources, the service desk and incident handling. The middle layers contain the IT systems indirectly related to production. At the center is the IT responsible for supporting the production process. Which system belongs to which category needs to be determined in detail with the customer and an inspector (pharmaceutical industry auditor), because producing and storing verification documentation becomes significantly more expensive as the categories move closer to the center.

IT systems in the simplest category are regarded as a black box or an **appliance** and do not have to be documented separately in accordance with the pharmaceutical company's regulations. The more systems fall into this category, the lower the cost for additional verification documents. It goes without saying that the IT service providers must also ensure that their own systems are properly documented according to the rules and regulations that apply to their own industry.

The IT systems in the highest category must be documented according to pharmaceutical industry requirements and included in its change processes and audits, because the regulations that apply to the pharmaceutical industry are stricter than those that apply to an ICT provider. The customer and the ICT provider must acknowledge and record regular changes to the systems. Moreover, these critical systems should be set up within an environment dedicated to the customer and not be operated as shared systems with other customers of the provider. Otherwise, any changes affecting all customers' IT systems must be documented separately for the pharmaceutical customer, leading to additional and unnecessary costs. Data privacy requirements can make this even more complicated, and the additional effort and cost of compliance can nullify the savings or efficiencies of shared systems.

The conclusion is that it is worthwhile to agree on a pragmatic implementation of sector-specific legislation with the customer. Different industries have different rules on, for example, whether and for how long each individual verification document should be saved and retained, or whether documentation and logs must be signed, or whether just storing these within an IT system is sufficient. It should also be obvious that, in the above example, the customer in a shared environment should not be able to see any changes made by other customers. Here, data protection takes precedence.

11.2 Near- and Offshore Regulations

Apart from increasing the efficiency of specific areas of the business, cost reduction is an important factor for many companies who consider outsourcing. Location and personnel costs are factors that can make the nearshore option an appealing one. Issues such as data privacy and data sensitivity should therefore be clarified at the proposal stage in order to determine which services can be operated nearshore or even offshore. If these issues are not discussed at the outset, there is bound to be a "rude awakening" – during the transition phase at the latest. Many unclarified questions about nearshore operation have led to IT providers breaching their contracts with the customers or have subsequently led to substantially higher costs when providers have been obliged to bring services back from nearshore. For this reason, we recommend that you maintain a compliance reference database that clearly shows the contractual arrangements. This will enable you to quickly refer to details of the nearshore arrangements made with the customer, data privacy requirements, and regulations pertaining to the operating sites and documentation.

11.3 Audits

It is preferable for audit preparation to be integrated with other tasks. The ideal situation is having all the necessary documentation available at all times, and for compliance to form an integral part of the business process. You can achieve this by scrupulously maintaining your verification documentation, undertaking regular internal audits, and by raising the awareness of employees of the issue.

11.3.1 Scrupulous Documentation

Regular verification documentation and record-keeping are essential because this demonstrates that important issues don't just exist in PowerPoint presentations or for the auditor. Business processes such as configuration management, change management and order management must go hand-in-hand with verification management and documentation. The better these processes are organized, the more supporting documents will be available for audits and compliance audits. If you take this matter seriously, you will reduce the amount of work needed for each audit and save the company money. Regular training courses should instill in employees the importance of careful documentation and the benefits it brings.

11.3.2 Regular Audits

We recommend that ICT providers carry out regular audits for their customers in addition to the obligatory checks required by ISO and ISAE 3402. At T-Systems, the latter is carried out by the independent public accounting firm Ernst & Young, who regularly certifies that all the controls are being applied and are working properly. Strategic audits should also take place. This means that whenever problem management encounters a vulnerability caused by a process, an audit should be carried out to examine it. The loss of power in a data center would be a typical example. In this case, the question to be asked is: what caused it? Because the regulations specify a back-up power supply. Other questions that the audit would need to ask are: does the data center have a redundant infrastructure? Were both power supplies operational? Would the UPS work? How would the applications deal with the switchover? How often is testing carried out? Based on the evidence and findings from this audit, the existing processes will then be modified where needed. As a result, the organization stays fit, and weak points can be targeted and eliminated quickly. One tried and tested method is to link audits directly to problem management. The principle should be: "If a critical error has occurred or an ongoing problem has been detected, an audit must be carried out."

11.3.3 Training and Raising the Awareness of Employees

Staff are often unaware that audits are not just internal to their company, but that they must be carried out in the interests of the customers. This is because customers also have to regularly demonstrate their compliance with the relevant regulations. Training should raise awareness of this, and should focus in particular on looking at the issue from the customer's perspective.

Moreover, staff need to understand the value of having verification documents available at all times. Each audit that has to be repeated because of missing verification documents, costs the company, as the external auditor has to be paid every time. This is particularly true for an ISAE 3402 audit, which requires the auditor to carry out more extensive checks. Any irregularities ("non-compliance") caused by non-existent verification documents are recorded in an audit report and provided to the customer. Too many of these irregularities will harm an IT provider's reputation – and in a worst-case scenario, customers could threaten to terminate their contracts, or even leave. Training should therefore also explain that preparations for an audit require significantly more work if the verification documents have to be produced specifically for that audit.

11.4 The Obligations of the Customer

The staff of the IT provider need to be made aware of their legal obligations – but so does the customer. Just because you outsource your IT does not mean that your IT provider becomes responsible for all of your legal obligations. A good service delivery manager will therefore regularly advise customers of their obligations. This applies in particular to security issues. A single customer who fails to take his security requirements and obligations seriously can have a major impact on the IT provider's other customers. Security flaws do not respect corporate boundaries. For example, experience has shown that standard passwords are still used too often out of convenience.

To be on the safe side, the provider can support his customers by regularly discussing verification documents and audits with them. When doing this, it is advisable to use a standard agenda and to be aware of which logs and reports the customer receives. To minimize cost, it also makes sense to use standard reports. The customer will then be informed regularly about the existence and relevance of verification documents and will be able to use that information in his own internal audits.

11.5   Industry Standards

Industry standards such as ESARIS (Enterprise Security Architecture for Reliable IT Services), for example, are very helpful in generating structured documentation. ESARIS is a security architecture developed by T-Systems for protecting complex production landscapes. It contains standardized instructions for all process steps and also specifies all the measures required to maximize IT security. These measures can be adjusted and supplemented in accordance with the specific needs of the company or business sector. For example, the integrity and availability of the IT systems of listed companies have to be compliant with SOX requirements and the processing of credit card data must be secure. The key benefit of ESARIS is its visibility. The information is available not only to the provider's employees, but the customer can also access it and monitor the progress of the measures introduced to secure their sensitive data. This transparency makes it possible not only to objectively evaluate improvement measures, but also to adopt a structured approach to thinking about verification of security, risk assessments and the conscious acceptance of risks.

Ensuring Maximum Security

Following on from our detailed look at compliance, we can now address another matter that is a vital part of any set of measures and initiatives designed to improve quality: security. For without security there can be no quality. **Quality relies on having effective security measures in place. However, a high level of ICT security is only possible if the quality is right.** This is truer today than it ever has been, as companies of all sizes and in all industries are being forced to tackle the problem of ever escalating security demands. A large part of this is due to the increasing digitization of business and society: the more things and processes are interconnected, the more opportunities become available to attackers. This is bound to increase in the future. Detecting security vulnerabilities and preventing hackers from accessing corporate information has become a daily struggle. Viruses, Trojans and similar threats don't take coffee breaks. ICT security must therefore be based on continuous processes and measures – and be viewed holistically. This includes a host of individual measures so numerous that there is not enough space in these pages to mention them all. As far as Zero Outage is concerned, we focus on three key aspects that are particularly relevant:

- **Closing security holes:** Security must always reflect the current situation. Security vulnerabilities must be closed as soon as they are detected and preventative measures should be in place to detect unknown ones.
- **Sensitizing employees:** Every computer and every smartphone provides a potential entry point for malicious code. Infection is made easier if employees do not know how to deal with the issue and are unaware of its seriousness. It is therefore essential to raise employees' awareness of the need for ICT security.
- **Institutionalizing responsiveness:** 100 percent security does not exist in the world of IT. That would be utopia. There is always going to be some element of risk. This makes it all the more important for companies to be prepared by having clear guidelines and command structures in place.

12.1 Closing Security Holes

Unfortunately, there seems to be no limit to the imagination – and the capabilities – of Internet criminals' intent on getting their hands on internal company information and confidential data. The statistics are frightening. Major network providers experience at least one million attempted Internet attacks each month in the form of port scans and DDOS attacks (see Kaspersky Security Bulletin 2013/2014). In addition to this, phishing emails regularly appear in employees' mailboxes. Social engineering – trying to influence specific people to carry out specific actions, such as disclosing information – has been very popular for a number of years. In addition to emails, other channels such as chat platforms, Twitter and YouTube can be used to disseminate phishing links. If a reader clicks on a link he has received, the attacker will exploit security vulnerabilities in his browser or in other software in order to gain access to data. Targeted and well organized attacks on companies are on the increase. Industrial espionage is the name of the game. Statistics from the manufacturers of anti-virus programs reveal that in 2013 alone, there were at least 1.8 million malicious and potentially undesirable programs in circulation. It is also known that 45 percent of the web attacks blocked in 2013 were carried out using malicious web resources located in the United States or Russia (see Kaspersky 2013).

The "Red October" attacks clearly show the reach of cybercrime and digital industrial espionage. The malicious software used in these attacks, which were directed at the computer networks of government organizations, energy companies, and research and military facilities, proved to be particularly persistent. Thanks to the malware's multifunctional, modular design, it was able to compromise a number of different platforms – computer systems and smartphones from different manufacturers – and steal a variety of confidential documents. Detecting and removing the malware proved difficult. When the security specialists from Kaspersky tracked down "Red October" in mid-January 2013, it had already been active, but undetected for several years. The spyware "MiniDuke" had similar intentions and was uncovered not long after Red October. The malware installed malicious code via Twitter and used Google as a backup for blocked Twitter accounts.

These are just some of the many examples and variants out there. These examples illustrate that **the methods used by cybercriminals are becoming more diverse and more sophisticated**. They keep perfecting their technology, optimizing their programs and improving their methods in order to get rich by stealing sensitive information. The lesson here is clear: it is essential to identify and close down vulnerabilities. It is negligent to adopt a wait-and-see attitude once a vulnerability has been discovered. There are many hackers and "freeloaders" waiting to take advantage of these vulnerabilities.

Quite apart from the exploitation of software vulnerabilities, **it is also surprisingly easy for criminals to gain entry to buildings to obtain information.** When a building is evacuated during a fire drill, for example, anyone with criminal intentions can steal equipment, documents and data and gain free access to computers that don't have a password-secured screen saver. Within the context Zero Outage, employees are made aware of any security vulnerabilities detected by auditors or security staff because they simply remove these unsecured devices. The employees then receive a friendly note telling them where to reclaim their equipment. This impresses upon them that they need to protect themselves and their company better.

12.2 Sensitizing Employees

Whenever there is a targeted attack on a company, you can be sure it has been planned. Some cybercriminals spend months preparing their attacks. They make a very close study of what company information is useful and what can also be stolen. And they explore many avenues in pursuit of their goal. They will examine the profiles of staff on social networks, for example, or create fake websites to mislead victims and infect their computers. This is why it is so important to sensitize the company's employees, because they are the ones who can make life easy or difficult for the attacker.

We often think of the theft of company data as a software hack. But there are also other dangers lying in wait for companies and their data. For example, some children were able to gain access to the ATM of a Canadian bank with the help of a handbook; they simply entered the default password it described (see T-Online 2014). Unfortunately, this is not an unusual case. You can always access a system if you have the necessary information. People prefer an easy life and do not change their passwords regularly, or they forget to change the password when initializing the system. Unless employees are sensitized to the potential risks, nothing will change. This applies just as much to clicking on unknown email content as it does to leaving confidential documents on the table instead of locking them up. When they are no longer needed, papers containing confidential information should be disposed of in the shredder rather than being tossed in the waste paper basket. Zero Outage sensitizes staff to these dangers and ensures that a basic understanding of security is part of every employee's security DNA.

If someone does happen to click on a malicious email or notices something suspicious, they should know whom to contact to report the incident. They should also be aware that they should not delete the suspicious email, so that the security team can uncover its source.

12.3 Exercises with the Customer

If, despite all security measures and rules, an unauthorized person succeeds in entering a system, a fast response time is critical for the system supplier and the customer. To make sure that all processes work properly when the worst happens, it is a good idea to run through emergency scenarios before they have a chance to occur and to regularly check response times. The more departments are involved in this, the better and more long-lived will the outcome be. Ideally, this kind of exercise will also be extended to include the customer and will take place as part of a "fire drill" in which the company checks its own quality assurance processes. **The efficiency of the alerting chain and the speed and focus of the interplay between customers, partners and IT will also be tested.** In other words: when an intruder is detected, who reports what to whom? Which areas of the organization are involved, when and with what information?

T-Systems regularly tests these processes with its customers, because these attacks can be directed at both of them at the same time. Test coordinators chosen from both sides work out a detailed attack scenario in preparation for the joint exercise. In doing so, they take into account attack patterns that are known or have taken place in the past. The test and simulated attack are then carried out over an entire day, becoming more and more intensive and more extensive as the day goes on.

These exercises should test:

- the effectiveness of the specialists' data exchange and teamwork,
- the coordination and communication between the teams on both sides,
- how quickly the crisis is tackled, and
- the quality of the decision-making.

Example Scenario

An employee has accidentally opened a phishing email and released a virus, the identity of which is unknown. The first warning was an unusually high level of inbound traffic and data being sent to an unknown IP address. How would you handle a case like this?

Step 1: The intrusion has been detected and an analysis of the IP address is carried out. As a result of suspicious processes, the customer's CERT (computer emergency response team) is now in close contact with our CERT.

Step 2: A search is carried out to find and quarantine the virus. In the meantime, the staff member has reported the incident to a security officer. The suspicious email is quarantined.

Now back to our scenario. The cyberattack continues. The attackers have succeeded in logging in to another of the customer's systems. This means they have been able to access commercial data and transfer it to their own systems.

If you are interested in reading about some real-life security scenarios that have happened, look up "Operation Night Dragon" on the Internet. You will find details of a fascinating case in which organized attackers from China had been spying on oil companies, probably for years. The intruders were able to steal data on oil and gas field production systems and production plants, and financial documents on extraction and bidding processes. Your search will turn up other cases of "security experts" who have used security holes in their customers' systems to blackmail them.

Step 3: The organization recognizes a crisis situation and involves all key stakeholders. A dedicated line is set up between the technical solution teams. A crisis management team and a crisis coordination team are ready. In this scenario, it has already been established that data has been transferred. The organization should now restrict the hackers' access by reducing the bandwidth of the ports being used to steal the data. Alternatively, the organization decides to disconnect the systems from the Internet entirely. One disadvantage of doing this is that it then becomes difficult to reconstruct exactly how the system was hacked, which means that there is still a risk that the hackers could get back into the network.

Exercise scenarios like this are also **carried out with the customer to see how quickly both organizations respond to events.** This involves testing how well the communication pathways operate and whether sensible decisions are reached without delay. The following questions can also be clarified:

- Are all the necessary people involved?
- How clear and accurate is the communication?
- Can the specialists get to work quickly and without interruption?
- Are the right experts networked and able to share their findings?
- Are the decision makers involved and kept fully informed about important developments?

Afterwards, a joint analysis of the process is carried out with the customer, including a critique of each other's actions. Based on this analysis, proposals can be developed for improving the way each party handles these situations.

As you can see, the exercise comprises a number of steps, each increasing in intensity as the crisis develops.

Another aspect that should not be underestimated is the importance of the role of **public relations** in security breaches. PR should be taken seriously and implemented professionally. If important public services are switched off, for example, the authorities and the population at large need to be informed in an appropriate manner. Also, in more difficult cases, the public prosecutor may have to become involved once the evidence has been secured. It is therefore worthwhile including this particular aspect in these exercises.

Business Continuity Management: Managing Crises Successfully

What service providers call business continuity management (BCM) their customers often call service continuity management: **avoiding unexpected disruptions and ensuring the availability of critical services at all times**. If disaster strikes, strategies and procedures should be in place to ensure that these services can still be accessed and used. The first question is: how exactly do you ensure the continued availability of key services?

Unfortunately, crises occur more frequently than we like to think. Which is why preparing and planning a Zero Outage program is so important. Typical examples of crisis scenarios might include heavy rain, flooding, fire, lightning strike, but also union action, epidemics and terror attacks. The challenge is to prepare for these events as effectively as possible. Deutsche Telekom, for example, prepares a detailed forecast of potential lightning strikes during the summer when there is a high risk of thunderstorms and uses this information to expedite the deployment of maintenance crews. Employees take part in regular fire drills and practice communicating with the local fire service in the event of a data center fire or other incidents. This involves, for example, letting the firefighters know which systems or circuits are critical for maintaining vital services and emergency networks so that, should the worst happen, they are aware that these must only be switched off in an extreme emergency.

Epidemics and union strike action have increased over the past few years. When they do occur, it should be clear which critical teams and sites are affected so that steps can be taken to supply replacement staff.

To prepare for these events within a Zero Outage context, you must be fully familiar with the services being provided and also be able to evaluate their importance. In addition, before practicing making decisions in crisis situations, you need to know which teams and which sites are essential.

13.1   Rating IT Criticality

One useful technique involves carrying out a risk assessment and classifying risks into groups. You need to determine which critical teams, services and sites exist and whether there is a replacement available for each one. The more critical the area, the less disruption or interruption is acceptable. There are, of course, some services whose partial reduction might be tolerable in a crisis situation. However, you must clarify how long you could do without these services, and what is "tolerable". **This should be based on the customers' critical landscape as well as an analysis and classification of their critical internal systems.**

The following should be looked at:

- IT server systems and applications run by the data centers
- Applications supporting the customers' business and support processes
- Essential data center infrastructures such as the power supply and air conditioning systems, as well as internal cabling, including basic network components such as routers and switches
- Desktops and back-end office communication systems
- Teams and groups of employees who operate these systems
- Sites at which these critical teams are located

If services are divided into classes, it is helpful to undertake a plausibility check in order to compare the costs of an outage with the cost of protecting against an outage.

You can then define priorities with reference to the high availability specification for mission-critical services and the acceptable downtime values for less critical services.

You are then able to specify, from the perspective of a **business continuity plan**, the maximum tolerable downtime for each application before it is recovered and brought back online. Bear in mind, however, that for technical and personnel availability reasons it will not be possible to restore all applications at the same time.

Example of allocation into classes:

- Class 0: high availability – no loss of service (for example, customer's critical production systems)
- Class 1: several minutes to two hours of downtime (infrastructure, service desk, technical service and back-up)
- Class 2: two to 24 hours of downtime (support systems such as HR systems, time tracking or intranet)
- Class 3: several days of downtime (post-processing systems)

13.2   Ensuring Staff Availability

There are many reasons why employees might be absent or unable to get to work. Consider, for example, a force majeure event such as road flooding, a major outbreak of flu, a bomb threat leading to the evacuation of buildings for several hours, police or military actions. Strikes are another possibility, although usually there is adequate warning of a strike to be able to prepare for it in advance. These events illustrate the importance of **safeguarding the workforce and the work they carry out**. Even more important is having a plan in place that covers all eventualities and specifies how to deal with each particular emergency situation.

13.2.1 Virtualizing Jobs

Is the building accessible? A network connection and employees working remotely, whether from home, a hotel or a café, could keep the work processes running. A solution that allows secure access to the corporate infrastructure via public networks would have to be established in advance. When doing so, you should clearly define which corporate applications may or may not be accessed externally. You will need to discuss with customers what access rules exist for their systems and applications.

A BCM or crisis situation requires clear communication with employees. This means specifying in advance who has the authority to declare a crisis and what form of communication is to be used to advise employees to either stay at home or travel to a network access point. The selection of communication channels is part of the detailed plan. These channels should also be tested and reviewed at regular intervals (about once or twice a year).

13.2.2 Managing the Simultaneous Loss of a Large Number of Employees

This scenario occurs during epidemics as well as strikes, and is covered in BCM planning. To prepare a suitable plan, you need to ask the kind of questions typically used in the preparation of a risk assessment, and base the BCM plan on the answers you obtain. Questions such as:

- Which services are mission-critical?
- To what degree might the performance of a mission-critical service be impaired?
- Which employees/teams are responsible for maintaining critical services?
- Where are the critical teams located?
- Can other staff temporarily carry out the duties of these teams?
- Must these duties be carried out in the same country or can they be transferred nearshore?

The last question is very important, because transferring work to nearshore employees could be a solution in the event of an epidemic or strike. These employees could then be given remote access to services, for example run the service desk, or work on changes in distributed teams. The obvious benefit of **distributing work nearshore and onshore** – to different geographical locations and countries – is that you will have greater flexibility when re-allocating work.

Replacing a team of world-class experts during a strike might not be so easy, however. This scenario should therefore be discussed in advance to find an alternative solution. It might not be possible to transfer some kinds of work abroad in an emergency. If so, it would be necessary to have an external team ready to take over. This would not be unusual in some highly specialist or top secret government departments, for example. In such cases, it would be advisable to form a partnership with another service provider who is prepared to regularly train its staff to take over in an emergency.

Plans should be prepared and stored for all possible scenarios. In practical terms, this means having backup teams ready and able to be deployed. You will also need clear rules about when work tasks should be transferred and how long the transfer should take. After a strike is announced, you would usually have two days during which to discuss this with your team. The same plans and considerations can also be used during an epidemic – although it not as easy to plan for situations where a near-immediate response is called for. This is why regularly testing and validating your plans is highly recommended.

13.3 Carrying out Regular Tests

BCM plans should be tested in preparation for an emergency. Useful tests include:
- Swapping work tasks, or job rotation, between teams
- Carrying out annual spot checks on the plans for transferring work tasks
- Conducting an emergency drill in which work tasks are transferred
- Regularly checking the teams' IT system permissions and access rights

Finally, you will find it useful to carry out smaller, regular trials with the people involved. All it takes to find out whether each person knows what to do in an emergency are a few impromptu phone calls and some brief, structured interviews. Training or briefing sessions can then take place based on the results of the interviews.

Partner and Supplier Management: Achieving Success Together

14

A chain is only as strong as its weakest link. This old saying applies everywhere – even in telecommunications and IT. And in these areas, different parties tend to be involved **because no telecommunications or IT provider can offer a complex service with all subcomponents from its own in-house production**. Thus, this book has presented mostly internal measures designed to increase customer satisfaction. These show what can be done within a company to prevent, or at least minimize, errors in projects and during operation. When a service disruption occurs, who caused the disruption is irrelevant for customers – and they are unlikely to find out. This is because the reason for an operational disruption may rest with a supplier or partner or with the ICT provider itself. For the sake of simplicity, from now on we will talk about partners and suppliers in general terms, without further differentiating between hardware and software suppliers or service and access providers.

Typical triggers of serious operational disruptions in ICT include defective network components that are not identified in the monitoring and prevent or delay failover to an alternative device ("flapping"). So, what should be done when a company's own production rate is falling while its dependence on third parties continues to rise? Increasing networking efforts with partners is a logical consequence of the market's development, because to respond to customers' requirements faster and better, it is necessary to create synergies and join forces with the best of the best. That said, companies must be able to count on their partners putting the same emphasis as they do on quality. The ICT service organization itself should therefore **have clearly defined quality criteria in place and communicate these to its partners and suppliers**. Such criteria may, for example, include only using material and software solutions that are particularly error-resistant, or the supplier's service processes being regularly audited in accordance with internal standards. At the supplier, responsibilities and escalation steps must be clearly defined, accessibility must be guaranteed whenever incidents occur, and critical services and service chains must be designed in a redundant manner. Standardized processes that apply to all partners in equal measure can help reduce the complexity of such an approach (see Kasulke 2013b).

T-Systems has communicated its own standards of quality to all partners and suppliers to ensure an equally high level of quality on all sides. In the next section, we present a **Zero Outage partner program** which enables potential external sources of interference to be reduced step by step and the zero-error principle to be communicated to all stakeholders. The last section then provides a brief look at a multi-disciplinary approach that we believe will help to better prepare the world as a whole to cope with outages in telecommunications and IT.

14.1 Taking Partners and Suppliers up on Their Promise

Let's start with the most obvious point: for most pieces of hardware and every software package there is a support agreement in place with the manufacturer. By contrast, in the open source sector there are no manufacturers because the software is developed by a community, but there are professional suppliers who update the public source code and release it with third-level support. Why, then, is the support of partners and suppliers the focus of our attention at all? In our experience, paid, guaranteed support is frequently not sufficient or is not provided quickly and decisively enough. Many technicians try to solve problems themselves because they have found that in previous cases the third-party support staff try to block inquiries by quoting additional costs, unclear terms of the support agreement, the wrong patch level of the firmware, etc. Technicians often find it faster and easier to plow through the FAQs, forums, manuals and release notes themselves than to draw on the contracted support. This, however, leads to the troubleshooting taking much longer because the company technicians do not have the same level of expertise as those from the manufacturer or provider of open source support.

To achieve good **collaboration between a company's technical teams and the external support teams** when using the **Zero Outage approach** we recommend escalating directly to the supplier's management (possibly even to top management) if a specific level of criticality is exceeded. This requires incident tickets to be opened at the relevant suppliers of all key components, based on standard actions in the incident process. This enables suppliers to requisition analyses and proposed solutions and request additional support from international specialists. The major advantage is that the partner can give feedback on whether it considers that the architecture of the solution, the qualifications of the company's staff and the entire setup are suitable for delivering a robust, high-quality service.

14.2 Zero Outage Compliance

Secondly, T-Systems has defined **Zero Outage partner and supplier compliance**, which has been deployed across all of its main partners. What does this compliance mean? The focus is on the three core processes of change management, incident management and problem management. For these T-Systems defined rules for cooperation with manufacturers. These rules are agreed with each manufacturer and tested in regular exercises. After demonstrating successful implementation of the rules and a positive test result, the supplier is then certified as "Zero Outage compliant" for one year, which forms the basis for successful negotiations in procurement.

The actual quality of the partners must be reviewed regularly. For this a dashboard should be drawn up that evaluates the most important factors, for example how often escalations involving the individual partners occurred, how the company teams subjectively view the collaboration, which recovery time was achieved in an incident, and if delivery deadlines were missed. In the context of Zero Outage, this dashboard is the basis for regular dialogue at senior management level in both companies and the foundation for any improvement programs and for changes in procurement policy.

14.2.1 In Incident Management

In incident management, everything revolves around rapid resolution of serious incidents – best measured using the mean time to repair (MTTR). To exercise positive influence over the MTTR, the general procedure described in section 14.1 should be used; a process should be implemented in which it is compulsory to open a supplier ticket for certain severities of incident. Based on this, the escalation process is defined together with the partner. This documents in detail through which channel (for example, using the "RedPhone") at which stages which management level should be involved, which support times are agreed with which criticality, and how communication between the parties involved is managed – specifically, when management conference calls take place, which reports must be prepared when, etc.

14.2.2 In Problem Management

In problem management, especially when the cause of serious incidents is unclear, it is extremely important to continue to work on analyzing the cause without interruption and with the best staff available, even after the service has been restored. Discussion about contractual penalties aside, the main focus is on gaining a rapid understanding of the causes to prevent the problem from recurring. For certain risk categories of problem tickets, T-Systems has therefore agreed that in the context of Zero Outage, problem tickets will continue to be worked on 24/7 where needed, and the partners concerned will continue to provide technical expertise and management support. In incident management, the cross-company process is activated after the service has been restored.

14.2.3 In Change Management

As described earlier, good change management is the most important lever for the prevention of incidents. For this reason, as part of its Zero Outage program, T-Systems also entered into agreements with partners. This is aimed at enhancing quality in the preparation and testing of changes and also ensures optimum support for the implementation of critical changes.

When partners are involved in critical changes they check the implementation runbook and provide feedback on the implementation strategy and the hardware and software used. During implementation, the partner makes technical resources available (including back-ups) and provides technical and management support as soon as problems arise. These measures can substantially reduce the risk, even in the case of complex changes.

14.3   Outlook: Zero Outage as an Industry Standard

Based on considerations from previous sections and practical experience, we are certain of one thing: to offer a top-notch service it is not enough for the integrator to solve his internal problems in relation to processes, technologies, and people. Normally, more than 20 different companies are involved in the creation of a standard SAP solution, from storage to the network and the server to the client, and a single failure anywhere is enough to disrupt a system's availability.

This is why we are pursuing one goal in our Zero Outage endeavor: we want to define certain standards in general terms and therefore **standardize processes, technical standards, and also levels of education across company boundaries** so that the risk of disruption and the recovery time are minimized. This is again about prevention on the one hand and, on the other hand, providing a rapid, resolute response when an incident occurs. The key success factors described in this book have thus been abstracted and principles have been derived from these on how the interaction of many companies can be improved to optimize overall availability.

The content of the standard can specifically be divided into four principal areas:

1. **Processes:** The participating partners undertake to keep relevant escalation points such as the "RedPhone" available for incident, problem and change management and to provide corresponding support to one another at the defined times.
2. **People:** Here, the focus is primarily on regular training and certification in key technologies in accordance with the standards of the participating companies. Integration providers undertake to certify their staff in accordance with the certification standards of the manufacturers, for example hardware and software manufacturers. Hardware and software suppliers undertake to certify their staff in the Zero Outage standards in incident, change and problem management.

3. **Technology:** General and specific criteria are defined to design, test and operate hardware and software at all levels robustly so that the risk is minimized. The most important thing here is a compatibility database across manufacturer boundaries showing which hardware and software from manufacturer A (storage, for example) is compatible with which hardware and software from manufacturer B (SAN switch, for example).
4. **Security:** Minimum rules are described to ensure that the availability of systems is not compromised by simple attacks and that vulnerabilities are regularly identified and eliminated.

This standardization enables pro-active prevention in addition to the reactive measures at process level. At present, 13 well-known companies such as IBM, Dell/EMC, SAP or NetApp have joined the Zero Outage Standard association (www.zero-outage.com) and are actively collaborating on the definition of this standard. The first version has been finished by the end of 2016, containing important initial standards e.g., for processes, mainly based on the principles described in this book. The plan is to update and extend this standard three times a year, focusing mainly on technology and security. It is important that this is an open standard, i.e., it is available to all integrators and suppliers and serves to improve cooperation between suppliers.

Successfully and permanently integrating a commitment to quality into a corporate culture requires this concept to be the focus of the company's value system. Every part of the company – without exception – is affected. The key is utilizing the human resources department as a hub for spreading the new culture throughout all of the other divisions. If quality is firmly anchored in the culture as a major criterion as early as the recruitment process, and additionally figures into compensation models, career planning and employee evaluations, the corresponding standards and values will cascade throughout the entire organization. The following chapter describes how a Zero Outage culture can be established and what is needed to do that.

15.1 Changing Behavior Patterns in the Organization

Have you ever thought about how children learn? Right: they copy behaviors, or assume the attitudes and values of their parents, people they trust, or role models.

To be honest, it is not different in the working world. We like to be led and guided by people and managers whom we value and trust. We all know the expression: **practice what you preach**. When managers are the embodiment of values and standards in practice and act as role models, their employees are more likely to adopt and associate these with this positive example. This approach is essential, particularly if you aim to hold your organization to high quality standards. Assuming that all **changes also cause friction and bring with them uncertainties and questions**, you will undoubtedly face resistance when introducing stricter standards, for instance in incident management and escalation situations.

If you want to transform an existing culture (possibly built up over many years and entrenched) into a Zero Outage culture, you must **set an example by starting with senior executives and top management**. "I have to get mixed up in operations myself?" you're thinking now. Yes, you should – especially because you are a manager.

In other words, here are a few tips on how to guarantee that your **transformation into a Zero Outage organization** will *not* succeed:

- You run your program out of the quality team, which is not particularly highly ranked in the company hierarchy.
- You are a lone wolf, and other departments are not required to pursue your quality goals. Your management colleagues will receive their bonuses even if they don't fight through resistance and leave their comfort zone to support your initiatives.
- Top management is not involved or only initially approved your strategy.
- You have not been requested to regularly present the topic to governing bodies and put it on the agenda. That means you can neither celebrate your successes nor push for cooperation, or force escalation.
- You are not involved in projects at points that require sustained engagement and attention to detail, such as training for new alarm chains in incident management.
- Your employees do not have the opportunity to learn from you how to effectively arrange a management or customer call, or that the most important parameters in the fight for Zero Outage are perseverance and discipline.
- When the phone rings at night as a result of a major incident, no one can be reached. After all, tomorrow is another day.

Admittedly these are extreme examples. However, please consider that your employees witness your actions day in and day out. They can see whether you really practice what you preach and whether you have the required discipline and sense of urgency in key situations, for example when a client's critical business service is affected. Do you drop everything for a critical outage at an important client (crisis mode)? Do you consistently orient your priorities toward finding a solution, and re-schedule all of your "urgent-but-not-important" commitments? No? Then how will your employees learn these desired behaviors?

Let's look at how the fight for better quality can really be won:
- Your company has defined Zero Outage as one of its top priorities for this year and in the coming years. There is a single mission – supported by everyone – and one strategy pursued by everyone.
- There are clear, measurable KPIs and a plan for what is going to be accomplished in the next 12 months, along with a set of long-term, strategic goals. Over-arching goals, such as reducing major incidents by x percent, are included in personal target agreements with senior executives and members of top management.
- In order to completely overhaul the company's culture, it is not the job of just one department to comply with the Zero Outage approach, but all of them, including those from which resistance is expected.
- A slot is reserved for a quality update in weekly management board meetings. The senior vice president (SVP) of quality provides an update on the most important quality KPIs, the highlights and low points of the preceding

week, and the key improvement programs. If necessary, decisions are made within these meetings. After that, action is taken and tracked to see if improvements are taking place – and the same is repeated again the next week.

- "Managers on Duty" are named from all IT operational departments and all management level employees. The managers are trained and are available according to a rotation schedule, even after normal business hours at nights and during weekends to lead the teams in the event of an outage, or, for example, to supervise significant changes during maintenance windows over a weekend. The contacts therefore have alternating responsibility for being available at any time of the day or night. And that is a critical point – because outages are not planned and can sometimes occur after business hours or during lunch breaks.
- In the case of a major incident, the manager on duty is the first to join the teleconference, and must push for and support the investigation into the cause until the failure is corrected. The Manager on Duty sets an example for a "sense of urgency" for all employees involved.
- Departmental and team managers are also responsible for quality in day-to-day operations. They have a checklist to use during daily quality checks with their teams that demonstrates how high the quality bar is set.

Of course, these are only a few examples of how mission Zero Outage can be woven permanently into all levels of the hierarchy. Every organization will have different drivers that you must define for yourself.

Afraid of Feedback? Establish a Feedback Culture!
The preceding chapters outline the fact that the success of Zero Outage hinges on discipline and high standards. Introducing these standards, or stepping on the gas, will also lead to changes and potentially resistance within the organization. For this reason, it is important that you remain **open to feedback from employees** from the beginning (notwithstanding mandatory minimum standards). Our practical experience shows that despite your best knowledge and belief, you might sometimes want too much of a good thing when it comes to mission Zero Outage – from the point of view of employees anyway – or it could be that certain details in the day-to-day operations at ground level were overlooked, etc.

Therefore, it is important to integrate key players from the operational departments into the development of your process standards and policies. If possible, assemble a virtual expanded management team with representatives from other departments who are multipliers, but who will also forward feedback from their employees to the quality team. Whenever possible, offer employees the opportunity to give feedback directly. This can be in the form of a question-and-answer session at the end of an employee call, an anonymous feedback survey with fields for free-form answers, or maybe business breakfasts in which you are available to employees to discuss and answer questions about the quality strategy in a relaxed atmosphere.

15.2 The New Ivy League of Employee Qualifications: The Quality Academy

In addition to the right mindset and the right attitude toward employees, a **compelling, up-to-date training program is an important component of the Zero Outage culture**. At T-Systems, we call this the Quality Academy. Why do we need an Academy? It's a think tank for the company-wide communication of all quality-relevant processes and IT training knowledge. Only employees who practice the Zero Outage culture daily and continually hone their skills can lead the organization to success.

In the past, T-Systems mainly ran training sessions on individual processes, for example incident management, which were rolled out in a "quality kit" in various versions once a year. The effort put into this was enormous: the rollout occurred at a defined point in time, occupied the entire organization for several months, and required labor-intensive tracking and reporting of the training sessions and certifications. At the same time, it was not possible to react quickly to the specific needs of individual departments with current information – particularly with regards to certifications. The point of the new Quality Academy was to build on the experience gained in previous years and also to optimize many aspects of the tried-and-tested training we offered.

The Philosophy behind the Quality Academy

We wanted to establish the Quality Academy as something new that would combine the concepts of "quality" and "academy" into a program of high-quality knowledge transfer. The name was also intended to convey that quality is important to us, and that both high quality and employee knowledge are core assets for our company. That's why we decided to create a dedicated academy specifically for this topic. Employees perceive this type of program to have a very different value than when quality issues are buried somewhere in a training catalog under 500 other choices – usually under "Other."

No Training without Certification

Once our continuing education vision had a name, the Quality Academy was launched on a unified training and certification platform in 2013. This can be a single entry point where all content is consolidated, a SharePoint, or an intranet page. It brings together training sessions from a wide variety of sources, and all employees receive a clear overview of the relevant options based on their role profile. Employees also have the opportunity to obtain a certificate after they have acquired the requisite knowledge in selected topics. This was an essential condition for the Academy: the point was not just to offer training sessions, but also to keep the knowledge level of our staff demonstrably current with the Zero Outage mission. On the one hand, certification provides important and useful proof of knowledge for employees. On the other hand, it is also a good indicator for managers, who get an overview of the knowledge in their teams. From the overall perspective of a global quality organization, it is vital for disseminating a broad base of knowledge and for rolling out new standards.

Advantages of a Centralized Training Platform

Unlike in previous years, there is no longer a limited time frame for the Quality Academy's certification phase, because the certification platform is available all-year-round. Certification is possible on a rolling basis depending on when employees completed their last training sessions. The certificates expire after 18 months and must be renewed, which ensures that employees are continually exposed to updated content. It also makes bringing new employees on board easier: they can complete training programs at any time and obtain certifications just like a "driver's license." It's an important way to experience success and also an ideal opportunity to embed the idea of quality and the corresponding expertise at an early stage.

Start Small: Core Processes and Proven Methods

Developing a Quality Academy takes a lot of effort. Some of this is devoted to motivating participating employees but more is devoted to the governing bodies that must be brought on board in the preliminary stages.

Do not underestimate the **effort involved in coordinating with internal stakeholders**:

- Human Resources: Continued professional training is an HR development matter and can be managed centrally for strategic purposes. HR will also provide you with information about job profiles, functions and training needs.
- Works Council: When you hear about "certification," you generally think of a training incentive, but the Works Council is occupied with the question of whether this might be an illegal form of performance monitoring.

At the same time, you want to ensure from the start that the training content is relevant to your employees' everyday lives in their functions and departments. You will also have to bring in key players from other departments and – in a global organization – from other countries. Initially, you will have to convey the fundamental idea to them and request resources. If you have no central budget, you will need support, even if that is just the willingness to participate in principle.

T-Systems has had good experiences with establishing the Quality Academy **high in the organizational hierarchy** from the beginning by presenting the idea to and receiving approval from top management. In addition, we developed the Academy hand in hand with Human Resources. This was also true of financing, which was centralized. Parts of the departmental strategic continuing education budgets were used to establish and bring to life the Quality Academy.

Since it takes approximately six to twelve months of coordination and development to launch a centrally placed Quality Academy, it makes sense to start with the issues most important to you, in order to quickly log some successes. **Prioritization** is key in this regard.

Here are some core topics for a Quality Academy you could start with:

- Incident management
- Problem management
- Change management

- Service asset and configuration management
- Order management
- Event management
- Plus integrating already existing IT and tool training relating to these processes

Additional training – for example, project management and general Zero Outage awareness – was intentionally developed only later in the process. The same applies to numerous other training sessions developed jointly with HR and the operating units for all service lines and countries.

The Next Step: Modularization

All training and certification opportunities are geared to specialist career paths. Our employees in operations, for example, are trained in the dual-control principle in change implementation. And our project management colleagues are interested in quality gates and touchpoints in projects. In our approach, each of them can choose three topics that most closely match their professional responsibilities and interests. The **individually designed certification questionnaire** contains 15 questions, five per selected topic. All three modules that the employees select are included on the certificate, which is valid for 18 months.

Another important point: if employees fail to complete their certification, **they cannot simply make repeated attempts as often as they want.** They cannot try again a second time until a certain period of time has passed, and the certification test is then based on a different set of questions. Each employee also receives different questions. This prevents employees from clicking through the test together during lunchtime. All questions answered incorrectly are listed after the test is scored, along with a reference to the training materials, a video tutorial, and a "CookBook" (process manual) in which the employee can investigate and find the correct answer.

Today, our training modules are even more modular than they were before. We record smaller segments of material individually and combine these into "playlists" to serve as training modules for various target groups. This makes us more flexible in updating the content and targeting particular groups more specifically. All employees receive training modules tailored to their specialist career paths and within the playlist can navigate from recording to recording – between chapters, essentially. That enables employees to **work through training materials faster while ensuring that the content is more specifically designed for each target group.** Employees who are only involved in a process or topic to a limited degree are therefore not required to work through a comprehensive training course, and instead only receive the content relevant to their responsibilities. This method also makes it easy to create training modules by combining topics that are universally relevant with special content for particular lines or countries. That supports simultaneous training on process and tool topics. In order to guarantee that the overall message is not lost, we also developed additional training modules that illustrate how various topics are connected into a bigger picture.

Modularization offers a number of other added benefits: usually, all of the chapters/content in a learning/training module do not have to be changed or updated at the same time. Some recordings, or learning modules, remain constant for longer periods of time, for instance those concerning process activities. Others change more frequently. Modularization helps to keep maintenance requirements as low as possible. Updating does not mean re-recording the entire training session lasting 60 to 90 minutes, but only selected content.

In short, the advantages of modularizing training content are as follows:

- Combined training on process and IT tool questions
- Flexible compilation of learning packages for various target groups
- Less effort required to update training content; faster updates
- Varied training sessions with different speakers for individual web-based training units

Growing Maturity: Playing with New Media and Formats

Once your content and structure is in place, and the Quality Academy concept has taken root, you can move on to the next step: teaching methods. Over time, we have changed not only the content of our training sessions but also our teaching methods. Attractive new formats, such as simulations, mobile training, and game-based learning provide variety in the usually uniform world of web-based training or recorded lectures with supporting PowerPoint slides. One method already established is the "flight simulator" or, as we refer to it, our **"driver's license" for the operations team**. This type of online training allows employees to realistically simulate and work through various scenarios and problems on their desktops to ensure that they are well prepared for real-life operations, thereby preventing human error. Only employees who have their "driver's license" are permitted to work on production systems. For example, we simulate the service management tool which employees must use to enter a change, including choosing the right configuration items (CIs), the correct group for testing and review, etc. They are not simply reading about the four-eyes principle in the process description but are trying it and going through the steps live in the system. This approach presents employees with an entirely different level of difficulty along with higher expectations for their ability to transfer knowledge. Without question, this type of format must be developed in conjunction with operations staff to ensure that their real needs are addressed.

Currently, we are also focusing on the **issue of sales enablement.** Have you ever thought about how your sales team talks about your company and the quality of your ICT products and services? Salespeople should know how to make an argument for quality, although they are themselves usually removed from operational issues. We organized the so called **"Zero Outage Awareness Training"** for this target group. In broad strokes, this training provides information about the quality program and focuses on the highlights also relevant for dialogue with customers, but does not go into greater detail. The next step is to develop game-based training on the topic of "How Do I Sell Effectively?" with a focus on quality. This game could involve fictional characters competing to solve problems or winning over customers with the right arguments for the Zero Outage approach in challenging situations that refer to quality.

Evaluating the Success of Your Quality Academy and Training Activities

You have probably already guessed that measurement is important for a Quality Academy as well. How often were specific training courses accessed? How many certifications were obtained as a result? What share of total employees participated in training?

In the section that follows, we outline a few of the key performance indicators we think are important for you to track to evaluate the success and **influence of your quality training program**:

- The scope of your Academy in number of employees (How many employees must in theory be certified and know the content?)
- Number/share of employees who completed the relevant training (Target: at least 80 percent for the year as a whole; the remainder accounted for by vacation, illness, turnover, etc.)
- Number/share of employees who were subsequently certified (Indication of the necessary support and also feedback on the difficulty level of the certification)
- Number/share of managers who completed the relevant training and were certified (Target in the interest of setting a good example: 100 percent!)
- Individual evaluations of the training courses to determine their usefulness and relevance to day-to-day activities (Short feedback survey at the end of each training session; indication of necessary content adjustments)

Additions can be made to the list above depending on the vision, mission, and maturity of the relevant program; it will change during the life cycle of your Quality Academy.

15.3 Communicating Zero Outage

Imagine that you have a **top-performing quality program, but no one is talking about it**. A very important factor in transitioning an organization to a Zero Outage culture is the issue of communication. In addition to external communication with partners and customers, internal communication must also be planned and implemented with particular care. It is one thing to implement standards, but another to keep them active and communicate them to employees in a timely manner.

15.3.1 Internally: Our Heart Beats Zero Outage

Assuming that you also communicate in your company via e-mail or a regular e-mail newsletter (push communication), the concept of "quality" should continually be a feature of communications within the company. The same is true for your company's intranet site.

- Is there a separate area or site where all quality activities and news are consolidated?

- Are the "heads" of the quality organization easy to find? Are contacts named? Is it easy to contact the quality team?
- Is all of your quality documentation (processes, rules, audit materials, guidelines, role descriptions, etc.) available in a single internal location? Is there a page that contains all of the links to other databases/sources so that employees can easily navigate this information?

An **intranet site devoted to ICT quality** and strategy should be judged according to the standard of successful e-recruiting and employer branding strategies: several current studies indicate that the most successful companies are the ones that link their "career" section directly to the start page of their website, where it can be found easily. Don't make the mistake of burying the "quality" section on the seventh level of the intranet next to the menu for the company cafeteria. It's true that when a topic is **visible, easy to find, and occupies a prominent position in the hierarchy**, this increases the "perceived importance" of the issue among employees, not to mention the click rate. If you are not sure, measure it. An analysis of clicks on new content or on the section in general should be in healthy proportion to the number of employees for whom this content – in terms of their roles – is relevant to their day-to-day work.

The easiest way to **communicate important content according to the push principle** has been and continues to be e-mail, particularly if it is sent regularly, perhaps as monthly "quality news." If you do not appear to have sufficient content for a monthly issue, the quality highlights can, of course, be published quarterly or as necessary, for instance if important milestones are reached or as an annual year-end review. In our experience, however, using several channels is a good strategy for disseminating content as quickly as possible. Each employee can then individually access and process the information when they have time during their workday. An **intranet is a classic pull medium** that must be actively sought out by the employee. This assumes that employees already have a specific need or that the page is already relevant to them because they know that the latest versions of the documentation can be found there.

Important information for which you want **immediate acknowledgment** should be "pushed," for instance through an e-mail directly from the process owner or quality officer. These e-mails can also just refer to the topic as a small teaser. A reference to the intranet or the company's social network can then offer additional, more detailed information, thereby automatically increasing the number of times these sites are accessed. This is a good option for raising awareness, particularly when you are building a new site.

Here are some recommendations for topics that can be communicated, along with suggestions for a suitable medium:

1. **New employees in key roles/leadership roles in the quality organization:** E-mail directly from the quality officer to all employees in the quality organization. If possible, a link to the intranet with an interview with the new colleague ("Five Questions for XY").
2. **Changes in core processes:** E-mail directly from the process owner to all employees involved in the

process and key suppliers/customers in the process. Short summary of key points, and a reference and direct link to more in-depth information in the process database.

3. **Unveiling of the quality strategy for the coming year:**
Invitation to all employees in the quality organization to an employee call, which might be held multiple times with the same content, but at different times of the day so that employees abroad in different time zones can also take part. Presentation in a web conference and recording of the audio and/or video of the entire conference. Making the recording available in the company's social network or intranet so that the information is also available to colleagues who were unable to attend.

4. **General quality highlights (for example, obtaining an important certification, contract signed because of excellent quality, summary of highlights for the year, accomplishments by the quality team, etc.):**
E-mail directly from the quality officer to employees (especially in other divisions), placement of the topic in the newsletters of other divisions, posting of successes on the company's social network and, if relevant to the public, on the company's homepage.

In all of these examples, the following is true: **you need to remove the hurdles that prevent your employees from accessing this information.** Employees are very likely to read an e-mail from their own company and especially from the management team if it appears directly in their e-mail inbox. It is important to link effectively to more in-depth information and remove barriers to access. Have you started a quality community in your company's social network? Make sure a link to this page is always included in the signature file of your newsletter. Do you want employees to look at what's new in a process description? Link directly to the corresponding document or directory. Do you want to invite them to an employee call? Provide direct access to Outlook's scheduling functions so that the teleconference can be double-clicked and saved immediately.

From a Chore to a Pleasure – Using Today's Media

Some companies even have their own internal TV networks. This channel should also be used to spread the word about the issue of ICT quality and its importance through messages from the board or management team. To build pride among your colleagues, you could film a video about quality and the most important "quality messengers," i.e., the employees who are actively driving the process. If you involve the corporate image/design department of the company, this can also be used for external communication, for example on the company's website. It doesn't have to be a major production with a five-digit budget. You could also use photos taken by your employees to make a video collage. Ask the management board member responsible for your program or the quality manager to say a few words about your company's ICT quality and highlight its importance for your clients. A well-made, emotionally impactful video can say a lot in 90 seconds – it's the best possible elevator pitch for your Zero Outage mission!

15.3.2 Suppliers and Partners: Strong Together

As a modern IT outsourcing partner, regardless of the specific field, every company is challenged by the fact that there is hardly a service left today that can be provided end-to-end without the help of other partners or suppliers. It doesn't matter if it's specialized software to be purchased or classic routers in a network providing mission-critical services to customers. Clients know this too, since they are also increasingly operating in a multi-provider environment, not least to avoid being too dependent on a single supplier. Your customers, particularly if you serve as an integrator, will expect a clear strategy outlining how you will guarantee the end-to-end quality of their digital services behind the data center's doors.

Against this backdrop, it is **imperative that you integrate your top suppliers and partners into your quality strategy**. To do so, you must define and answer questions such as:

- Is there a joint quality strategy, or are you handing down your Zero Outage mission to your partners and suppliers?
- What benefit does the joint Zero Outage strategy have for your joint clients?
- How do factors such as redundant technologies, clearly defined processes for all partners, and highly qualified staff increase the availability of the ICT services offered along the supply chain?
- And if it is already too late to lock the stable door: in the case of an outage, how will you work with your partners and suppliers to ensure fast, end-to-end restoration of services? How does the alarm chain work?
- What are the successes to date in your joint initiative for better quality?

Beyond this, you should define the expectations you have for your partners and for their fulfillment of your quality standards – and **how the partnership benefits from that.** This assumes the following:

- Your partners share your understanding of a zero-error culture and make suggestions on how to improve quality continually and for the future. Both sides profit from this.
- Your partners are committed to high quality standards exceeding standard SLAs: that could be Manager-on-Duty support within x minutes after an outage begins, a maximum resolution time of four hours, or active support for problem and change management.
- Your partners and suppliers regularly have their staff undergo training: they ensure that their own employees are as familiar with your quality standards as yours. Training in incident, problem, and change management is included in this along with a continual exchange of information regarding best-practice solutions.
- Your partners have their own services certified according to your Zero Outage standards. After the initial certification, a follow-up review is held once a year.

It is then very easy to **communicate this value added to your partners and suppliers**:
- Expertise:
  - Variety of options for cooperation
  - Best-practice sharing: learning from the previous successes of your Zero Outage program
  - Sustainable optimization of their own products and processes (greater efficiency, etc.)
- Budget:
  - Promotion of proactive quality measures
  - Attractive opportunity for additional business through the joint development of services and solutions
- Brand:
  - Prominent Zero Outage branding (e.g., "Trusted Service Provider") in tenders and new business activities

Again, communication is very important when working with partners and suppliers. Ultimately, everyone has to work in concert. Both parties have the same goal: to ensure long-term customer satisfaction and loyalty with the best possible products and services. Partners and suppliers are a key factor in achieving end-to-end quality in your company.

15.3.3 Externally: Create Market and Customer Enthusiasm

Ultimately, even the best quality is not an end in itself. Regardless of your sector or industry, a company that offers premium-quality products or services will not succeed in the long run if these are not perceived as such by the market.

In addition, the whole point is not just to attract new customers but – and this is often overlooked – to keep existing customers. **How much do you discuss the issue of quality with your existing clients?**
- How many clients did you explicitly talk to about quality – both positive and negative – in the last financial year?
- Is the issue of quality of operations, projects, compliance with SLAs, etc. regularly (ideally monthly) on the agenda of your sales department's service review meetings or service management with your clients?
- Do you have an overview of all of the quality problems your customers have with your services?
- How do you process these in your organization? Who responds to the customer or gives status updates? When? How?
- Is there a joint plan with the client to resolve existing quality problems? Is this in turn reviewed (ideally monthly) in service review meetings?
- Do you celebrate milestones with your clients such as the launch of major projects, or successful transitions or transformations? Do you help your customers market success stories within their own organizations?

Quality and Zero Outage cannot truly be experienced until they catch on with your clients as well. In the end, the deciding factor is customer opinion, which is influenced both by the services provided and measured against the SLAs and by perceived quality. And, as described in Chapter 10, these two things can differ.

15.3.4 Zero Outage Communication Roadmap: Where is the Journey Taking Us?

Just like quality, the issues of culture and communication also cannot be rushed. Here again, a well-thought-out plan is the most sensible approach.

If you are starting from scratch, first work with your core team and, if necessary, experts from your communications department, to develop a **"Zero Outage Editorial and Communication Roadmap."** Effective communication is like fishing: the bait must appeal to the fish, not to the fisherman. So, make sure you identify **everyone with a stake in this issue**.

- Who are your internal and external quality stakeholders? (Quality team members and their supervisors; the management board; financial control, sales, and service management departments; operations and project management; customers; external analysts; sourcing advisors, etc.)
- Who provides input for your processes or expects your output? (Examples: process managers provide KPIs and corresponding analyses and prepare project proposals; management expects a regular overview, especially concerning problems and progress made on solving them; operating units require very specific analyses to implement measures)
- With whom are you "wrestling" for better quality? And whom would you like to address specifically for that reason?
- Who are important opinion leaders and "patrons" for an issue that you would like to promote, or who could serve as multipliers?
- Which stakeholders can you consolidate into groups because of their requirements for information overlap?

Because one form of communication will never be sufficient to reach all stakeholders, you must consider how you will reach various target groups. For example, you would communicate different information to employees than top management, and at different intervals. You must therefore also consider **the specific content of your communications, types of media and frequency.**

- Which topics and updates are of interest to which groups?
- Which information must be provided to which target groups without fail? (For example, request for employees to obtain annual internal quality certification)
- How has each group received information to date?
- Do you also want to involve the group to a greater extent or promote constructive discussion?

- Which media can you use to reach each group most effectively? Where are you getting the best response from? What media mix would be effective?

We recommend that you invest some time in this step and also personally survey representatives of your stakeholder groups. Talk to your employees and ask them for feedback. In this age of information overload, every reader will be asking, **"What's in it for me?"** That means that the more relevant your quality communication is for the recipients, the more likely it is to be read. And the better and more impactful it is (which usually means how brief), the faster and simpler it is for people to digest. Think about the many feedback surveys that you now receive in your free time. When you are shopping online, are you more likely to answer if the site asks, "Please answer three questions to give us feedback on our packaging" or "Please evaluate our logistics process. It will only take 20 minutes of your time"? And, since we are talking about feedback: **communication is not a one-way street.** From the start, you should think about how you will gauge your readers' opinions from time to time. You could send out a web-based feedback survey to follow up on employee calls, for instance.

The third and final step in creating your "Zero Outage Editorial Roadmap" is to write out specifically (even just in a simple Excel spreadsheet) the information that will be communicated and when, through which channels, who will supply the content each time (include run-up time in your planning), and what approvals must be obtained within your company, if any. Regularly update your editorial roadmap with your team, for instance every 14 days. On the one hand, that helps you maintain a cadence, but on the other hand, it also allows you to remain flexible and react quickly when new issues come up.

15.4 Culture-Promoting Factors: Do You Know Your Quality Stars?

In this section, we would like to present two ideas for internal campaigns which, in our case, ultimately caused breakthroughs in awareness about Zero Outage and promoted pride amongst our employees about a job well done. If you previously thought that marketing and branding were only necessary for selling products outside of the company, then this is our attempt to convince you otherwise.

Example 1: The "Quality Star Awards"
Three years after the launch of the Zero Outage program at T-Systems, the program up to that point had consisted mainly of the roll-out and audit of required standards – in other words, topics that did not trigger much positive response. Then, a member of our quality team had a lightbulb moment: she wanted to give back to employees for their hard work. Ultimately, the success of the entire program is built on the involvement of our employees, so the idea was to reward them for their commitment. The concept of an award was born. In order to make Zero Outage more than an ob-

ligation and responsibility for employees and to associate it with pride, T-Systems developed the "Quality Star Awards." The philosophy behind this competition was, and still is, very simple: employees and teams with outstanding quality achievements are recognized as Quality Stars. The Quality Star awards are held once a year. All nominations are documented centrally in the quality community in the company's social network. To generate the required interest in the awards, the start of the competition is always announced in a direct e-mail signed by management, for whom these awards are of special interest. The call is sent throughout the company to the three divisions and to all countries. After all, Quality Stars could be anywhere, even in corners of the organization where we may not expect to find them. The most important thing about these awards? Employees nominate themselves and their colleagues. The Quality Star Awards are meant to both promote a mass movement, and reflect an honest assessment of and recognition for what has been accomplished.

In order to do justice to several areas of focus, the quality department at T-Systems awards this honor in three categories, each with three places:

1.   Quality Idol (Individual award; for individuals and excellent quality role models)
2.   Quality Starlight Team (Team award focusing on operations)
3.   Best Zero Outage Initiative (Project award; for outstanding initiatives to promote and improve quality)

For approximately one month, nominations can be submitted or the nomination template can be uploaded to the quality community, after which the winners are crowned at the annual international quality management meeting. Over 100 nominations are now submitted each year, so a shortlist is narrowed down in advance based on specific criteria. Here are a few:

- Creativity of the application
- Completeness and information presented
- Value added and achievements clearly discernible for Zero Outage and quality.

After that, management steps in: the shortlist of nominations from all over the world is sent to the quality officers of all countries and units around two weeks in advance. Each quality officer is tasked with defending his or her nomination choice and making a 90-second pitch prior to the final voting to determine the winners. Why should the Mexican team win? What makes this project from Slovakia so special and unique? Why does employee X deserve to win the Quality Idol Award? The nominations are then put to a vote. The winners receive a monetary prize and are honored by management and in the newsletter, and their names are announced in company communications on various channels. The individual countries pick up the news in various ways, including in-depth interviews with the winning teams. The award is undoubtedly a great honor for each winner and also promotes the visibility of the issue.

At the first Quality Star Awards in 2014, management was overwhelmed by the positive response from employees. After just a short time, company-wide communications generated over 1,000 views in the community. After four weeks, nearly 100 applications had been received from all of the countries where we do business, which exceeded all expectations. The nominations were very creative and presented outstanding and sophisticated projects and initiatives for improving quality that management did not even know existed. Prior to this, the opinion prevailed that, because they originated in Germany, the awards would motivate colleagues in Germany more than others, and then only those from the quality department. But quite the opposite was true: the response from the EMEA, APAC, Americas, and Nearshore offices was much greater, and the concept was received very positively in those areas.

Employees were very motivated and developed creative campaigns that warmed the heart of every committed quality fan. In Malaysia, for example, our colleagues designed buttons with a saying about their quality commitment (see Fig. 15.1). They distributed these to their entire staff, and the head of quality for APAC, who wore one personally to the awards ceremony, also proudly presented a button to the German management team. Everyone loved it.

Fig. 15.1 "Quality Commitment" Buttons

Another great initiative came from Hungary. Our colleagues there developed new types of training to make conveying process knowledge more exciting and interactive. These "Process Awareness Days" were also successful. The Quality Star Awards were, and continue to be an incubator for the global Zero Outage mindset in the organization.

Example 2: "Our Heart Beats Zero Outage" Campaign

Another very successful internal campaign was run in 2015 under the title "Our Heart Beats Zero Outage." This campaign was designed to evoke emotions and revolved around a simple gesture: employees quickly tapping their chests twice – like a heartbeat. Based on this gesture, employees then used simple tools (their own smartphones if that's all they had) to film a short selfie video with no make-up or set, just in their offices, headquarters, or cafeteria and – above all – with no formal script.

All of these selfies were then compiled one after the other into an internal video that presented a fascinating cross-section of our workforce around the globe making the "Zero Outage gesture." Particularly well-made videos were used in the new version of the quality video intended for external target groups. This also makes it clear: employees make the difference – their hearts beat for Zero Outage (see Fig. 15.2).

The videos made were also launched centrally by the corporate communications department and made available to all employees to a very positive response. The main theme continues to be used, for example, on posters hanging in the various offices. In any case, the success of this campaign stemmed from its authenticity and from the use of real employees who live and breathe Zero Outage in their day-to-day work. Other employees could much easier identify with them than with actors cast for the part.

Fig. 15.2   "Our Heart Beats Zero Outage" – Employees' Selfie Videos

15.5 Making Zero Outage a Permanent Fixture in the Company: Be Patient!

The goal of all Zero Outage efforts is ultimately to weave this common vision permanently into the corporate culture. A major step towards this goal is **actively integrating Zero Outage into your organization's value system.** And this must positively impact employees as a result of concrete incentives. Possible approaches for embedding Zero Outage into the culture include acknowledging quality-promoting behavior in particular through raises or promotions, or by giving an award. This motivates employees to align their actions with a Zero Outage culture, thus ensuring that the desired understanding of quality is reinforced throughout the organization.

It goes without saying that initiatives like this **bring with them investments, both in terms of finances and staff.** In addition to adjusting HR development activities (such as training courses and evaluation standards) and recruiting strategies, it is crucial that management consistently acts as a role model over the long-term. The fact that quality is given priority must be supported through regular communication, thorough reviews and continual enhancements.

Summary

There is no stopping the digital transformation: according to a Bitkom survey, 72 percent of companies believe that this is one of the largest tasks currently facing them – second only to finding enough skilled staff (73 percent). And 40 percent of CEOs and board members are convinced that they have fallen behind "competitors who embraced digitalization early on" (Bitkom 2016). CIOs and other ICT officers must therefore strike a balance between guaranteeing reliable IT operations while also promoting the digital transformation of their own companies and developing innovative solutions. The only way to achieve both of these things is to make quality the top priority.

Quality is much more than just a hygiene factor. Quality is the "glue" of digitalization that holds everything together. Digital transformation can only be a success if the underlying information and communication technology is reliable. The business capabilities and thus the very existence of companies depends on it these days. But quality doesn't come out of nowhere. Ensuring quality in ICT is a complex management task. Countless components must interact smoothly at all times so that production or sales, for example, can work without disruption.

But how can this complexity be handled? How can IT officers ensure stable, reliable ICT operations? The answer: with standardization on all levels. For the maximum quality and reliability in ICT, clear standards are needed – for processes, for technical platforms and for employee training. These standards must not only be introduced and implemented, they must also be followed consistently. Central governance is therefore essential. This is the only way to effectively reduce the complexity of ICT, with all of its rapid new developments and quickly changing demands.

But quality is not only a question of clear standards, it is also a question of attitude. Human error is still the most common cause of IT disruptions. And the best training is worth nothing if employees don't pay enough attention to quality assurance. The only solution here is a comprehensive approach that systematically raises employee awareness of quality and ensures that everyone – from trainees to senior managers – is committed to a zero-error culture. A strong quality organization that is actively integrated in operations processes must consistently and persistently promote this Zero Outage principle on every level of a company.

And quality awareness must not stop at the boundaries of a company. Today, no product is produced exclusively by a single manufacturer. Companies of all sizes work together across different industries. This means that there are always more interfaces and more points of friction. If everyone isn't working with and maintaining the same high standard of quality, then the result can be faulty products and outages. Furthermore, the highest standards of quality are essential to innovations such as robot-supported medical procedures or even self-driving cars.

Smooth cooperation is only possible when everyone sees the bigger picture and shares the same standard of quality. The ICT industry needs a network of partners who are committed to the zero-error principle and follow shared rules for quality management. What standards should we follow to develop fail-safe products? How mature do new components need to be in critical systems? And what kind of reaction times during outages are we committed to? If we can jointly answer these and other questions, then customers worldwide will benefit from Zero Outage, or fail-safe ICT. And we will lay the foundations for the successful, sustainable digital transformation of industry as a whole.

Annex to Chapter 3: ISO, ITIL & Co. – A Baseline and Orientation How-To

Defined Standards

International Organization for Standardization (ISO)

ISO certifies the behavior of an organization and its compliance with the procedures defined in the standards. ISO is an industry-neutral, internationally recognized body.

Advantages of ISO certification:

- "Long-term quality assurance
- Discovery of potential for improvement and cost savings
- Increased customer and employee satisfaction
- Positive public image
- Risk minimization
- Greater cost effectiveness via process improvements
- Enhanced competitiveness
- Satisfaction of specific customer requirements" (TÜV NORD GROUP 2014)

ISO 9000

This standard provides companies with guidance on selecting and applying standards involved with quality management and quality assurance certification. ISO 9000 is not itself a certification: instead, it helps companies to identify the correct ISO standard, from 9001 to 9003. ISO 9000 therefore serves a guide to the subject-matter and terminology (see Glaap 1993).

ISO 9001

This standard is applied to certify quality management within a company. It is the most widespread standard in use domestically and internationally and is thus of critical importance when a company wishes to publicize its high standards of quality while simultaneously improving its efficiency. ISO 9001 is industry-neutral and is the strategic cornerstone for any company seeking long-term improvements to its quality management systems. Advantages for companies wishing to obtain certification to this standard are:

- External visibility of company-internal processes
- Increased customer satisfaction
- Lower error rates with associated cost reductions

ISO 9001 is based on eight Quality Management (QM) principles (see TÜV 2015):

- **Customer focus:** The company should carefully survey customer and market requirements, assess the extent to which these can be fulfilled internally, achieve the level of performance in accordance with specifications, and, lastly, determine customer satisfaction.
- **Leadership:** An ISO 9001-certified QM system provides top management with a management tool. It is the task of the management team to maintain and develop this tool. Top management must themselves play an active role, substantiating QM with a clear vision, mission statements and targets.
- **Involvement of people**: People at all levels are the essence of an organization, but staff can only truly develop their potential on the company's behalf if they are properly involved in company processes – and this involvement also works to boost their motivation, dedication and creativity.
- **Process approach:** An ISO 9001-certified QM system should represent the actual optimized operational processes. To obtain the desired results, all of the activities and the associated resources should be defined as processes and managed efficiently.
- **System approach to management:** A system consists of a network of processes, which interract with each other. Only if these interactions are understood and managed can the company achieve its goals effectively and efficiently. The system approach to management results in the structuring of processes while discovering the interdependencies between them.
- **Continual improvement:** Continual improvement in the context of ISO 9001 is essential for healthy business development. Successful companies respond to expectations from customers and the market, continuously optimizing their products, services and processes. A systematic culture of improvement fostered throughout the company can increase its potential performance and secure a competitive edge for the business.
- **Factual approach to decision-making:** Effective decision-making is based on analyses of information and data. Applying this principle supports decisions that are based in fact and on reliable data. Opinions and decisions can be compared and evaluated through facts, figures and data. Keeping records of the decision-making criteria used also permits retrospective assessments of the effectiveness of specific actions. For example: If a company has had x instances of IT outages in the last 12 months and expected quality measures to produce improvement by y, but results were not remotely like y, then other measures must be identified.

- **Mutually beneficial supplier relationships**: A company is dependent on its suppliers – and vice versa. This makes it especially important to establish a win-win situation that enhances the ability of both to create value. This QM principle strengthens trust between the company and the supplier, enabling a long-term, mutually beneficial working relationship (e.g., a long-term supplier agreement) to develop.

ISO 20000

While ISO 9001 provides industry-neutral certification for any business, ISO 20000 deals specifically with issues affecting the IT sector, providing a framework for efficient IT service management (see itwnet 2016). ISO 20000 builds on ISO 9000 – so is redundant in places – but caters specifically to the needs of IT companies. In this context, ISO 20000 sets out the associated minimum requirements for certification. The standard does not reiterate basic principles such as customer focus, leadership, etc. (see ISO 9001).

ISO 20000 consists of five parts (see Beims 2012):

- Part 1, "Service Management System Requirements": Contains all of the "shall" requirements that are necessary for certification.
- Part 2, "Code of Practice": Contains the "should" requirements plus guidance on implementing the methods from Part 1.
- Part 3, "Guidance on Scope Definition and Applicability of ISO/IEC 20000-1": Supplements the information in Part 2, focusing on the implementation of a service management system (SMS).
- Part 4, "Process Reference Model": Contains a process reference model for Service Management processes and provides advice on setting up a process assessment model.
- Part 5, "Exemplary Implementation Plan for ISO/IEC 20000-1": Contains guidance on implementing service management capable of passing certification.

ISO 27001

This standard governs IT security and its deployment in companies, local authorities and non-profit organizations.

"This standard is thus intended

- to formulate requirements and objectives for IT security;
- to promote the cost-effective management of security risks;
- to define management activities involved with information technology; and
- to ensure the fulfillment of goals specific to information security.

Contents [...] of [...] the standard:

- A description of requirements for management (production, rollout, operations, monitoring, maintenance and improvements)
- Caters to all types of organization (e.g., commercial businesses, government authorities, non-profit organizations)

- Implementation of appropriate security mechanisms by requirements specifications" (TÜV NORD AUSTRIA 2016)

IT Infrastructure Library (ITIL)

Service Strategy (SS)

It is important to understand that the IT service provider should be given autonomy within the company. Decisions, strategies and portfolio analyses let the service provider become an entrepreneur within the enterprise – and this is desirable. The provider must make autonomous decisions and should be accorded the rights to do so. Service strategy addresses the necessity for IT service providers to prepare for the eventuality that they can be undercut by other providers offering cheaper and/or more efficient services. Providers therefore need to continuously improve their competitiveness and scrutinize their own performance as an ongoing process. This is a crucial step – whether or not the provider is supplying external customers or "only" the rest of the company: even in an internal context, external providers will eventually appear, promoting their services as a more efficient solution.

IT service providers therefore need answers to some basic questions. Which services can and should they be offering? What can be produced cost-effectively, and with which products is the provider ahead of the competition? The company must also adopt the customer's point of view. To make this easier, ITIL 21 formulates various product benefits that improve the IT unit's understanding of the customer perspective. In this way, the basic premise of the working partnership shifts fundamentally. IT no longer waits for the customer's order and then completes it. Instead, the service provider becomes intimately familiar with customer processes, regularly re-assesses the portfolio and approaches customers proactively (see van Bon et al. 2010).

2.5.4.1 Service Design (SD)

Service design brings us to the specifics. Instead of considering all products in general, the focus is now on the specific solutions offered to customers, i.e., the product portfolio for individual clients. But service design must always remember that these are the same products that were specified in service strategy.

Another component of service design is the use of service level agreements (SLAs) (see van Bon et al. 2010). This topic is addressed separately in Chapter 10.

2.5.4.2 Service Transition (ST)

Typically, companies manage to complete changes in IT infrastructure or the rollout of new software, for example, only with considerable effort and the related risk that the productivity of the business will suffer massively during this period – especially as a result of errors or unforeseeable outages. What's more, the customer may suffer financial losses if business continuity cannot be assured. Even for relatively minor changes, this quickly becomes a stumbling block, and customers will often reject new software or software updates that could further their business. For the service provider, too, even smaller-scale software updates are critical, since the provider must always ensure that corporate IT is up-to-date to safeguard the customer's ability to do business in the future. This also works to promote customer loyalty and is an important factor in ensuring that customers are not ultimately lured away by a seemingly more innovative competitor. Crucially, it also signals that the IT partner is capable of independent service improvement and that the customer can rely on the service provider implicitly. Transition must therefore also be a self-contained unit – to drive continual optimization of this process and the establishment of best practices.

Within ITIL, service transition is subdivided into:

- Service Asset and Configuration Management (incl. the Configuration Management System, CMS)
- the Configuration Management Data Base (CMDB), in which the Configuration Items (CIs) are documented
- Knowledge Management (KM)

The objective of these areas within service transition is a high degree of specialization, the full documentation of expertise, and the critical assessment of the unit's own work by KPI monitoring and audits. In addition, ITIL also emphasizes the importance of training following software changes or the introduction of new software. This is not only intended to encourage scrutiny of the need for the updates but also to establish if the customer perspective is being adequately considered: what do employees think of the software and how can staff become familiar with the new solution as quickly as possible? (see van Bon et al. 2010)

Service Operation (SO)

Service operation aims to ensure that operations proceed as smoothly as possible after rollout, and covers issues such as handling customer queries, faults, backups, etc. While incident management is intended to prioritize resolving existing incidents as fast as possible, problem management handles the task of analyzing incidents – including recurring incidents. True to the concept of prevention, the focus is not merely on resolving a fault but identifying its root cause to prevent reoccurrence in the long term (see van Bon et al. 2010).

Continual Service Improvement (CSI)

This ITIL volume aims to raise awareness about achieving improvements in products, service levels, quality and processes on a permanent basis – although it should be stated that CSI is not to be viewed as a self-contained step at the end of a value chain but must run continuously within each and every phase in ITIL.

Contents of this volume:

- Processes (How do I initiate an improvement process? What are the key factors for service reporting, service level development and service measurement?)
- Staffing structure (How should key positions such as project manager, service manager, CSI manager, etc. be filled?)

- Tools and methods (Assessment, benchmarking, balance score card, etc.)
- Rollout to operations (What are the prerequisites for implementing CSI?) (see van Bon et al. 2010)

The 2011 version is based on ITIL V3. This defines 34 processes and eight or nine functions, which are intended to regulate the IT life cycle. The subject matter is very wide-ranging and here too it is up to the individual company to identify the processes most appropriate and necessary before applying them.

The processes below are actively practiced, in roughly this order/prevalence/maturity level:

1. Incident Management
2. Request Fulfillment
3. Event Management
4. Access Management
5. Service Level Management
6. Change Management
7. Problem Management
8. Configuration Management

And Service Desk is an actively practiced function.

Project Standards

PRINCE2
The Seven Processes in PRINCE2

1. Starting up a Project (SU)

This process comprises project preparation. The project manager and the customer/client coordinate their approach. The basis for this change is typically a business case. In addition, this preparatory stage is also used to set goals (and non-goals not in scope for this project), put together the project team and appoint the project steering committee. Steering committee members are typically chosen from project stakeholders.

At this stage, the to-do list consists of:

- SU1: Appointing a Project Executive and Project Manager
- SU2: Designing a Project Management Team
- SU3: Appointing a Project Management Team
- SU4: Preparing a Project Brief
- SU5: Defining a Project Approach
- SU6: Planning an Initiation Stage

2. Directing a Project (DP)

This is a process that runs parallel to the entire project and is intended to maximize the quality of work and ensure a high chance of success. The process involves monitoring the project and each individual stage.

To-do list:
- DP1: Authorizing Project Initiation
- DP2: Authorizing a Project
- DP3: Authorizing a Stage or Exception Plan
- DP4: Giving Ad Hoc Direction
- DP5: Confirming Project Closure

3. Initiating a Project (IP)

The project commences, project management processes are defined and a detailed plan is drawn up. A project result must also be defined. The core management product is the project steering documentation, which is created at this stage. Output is as follows:
- Quality plan
- Configuration management plan
- Project plan
- Communication plan

The to-do list for this stage is:
- IP1: Planning Quality
- IP2: Planning a Project
- IP3: Refining the Business Case and Risks
- IP4: Setting up Project Controls
- IP5: Setting up Project Files
- IP6: Assembling the Project Initiation Document (PID)

4. Controlling a Stage (CS)

This defines the project manager's day-to-day tasks, i.e., project control, planning and monitoring. Work orders are referred to as work packages. Package status (completed/not completed) provides information about the project's overall status. It is important to define measurable goals for the packages and thus obtain an abstract metric for project status, as measurability is otherwise difficult. The to-do list for this stage is:
- CS1: Authorizing a Work Package
- CS2: Assessing Progress
- CS3: Capturing Project Issues
- CS4: Examining Project Issues
- CS5: Reviewing Stage Status
- CS6: Reporting Highlights
- CS7: Taking Corrective Action
- CS8: Escalating Project Issues
- CS9: Receiving Completed Work Packages

5. Managing Product Delivery (MP)

PRINCE2 defines the products created by projects as "management products". "This process creates the project's products and this is where the majority of project resources is deployed" (Ebel 2011).

In this stage, the work orders parceled into work packages are executed after having been planned and authorized in the preceding project stage(s).

To-do list:

- MP1: Accepting a Work Package
- MP2: Executing a Work Package
- MP3: Delivering a Work Package

6. Managing Stage Boundaries (SB)

At this stage, the intention is to permit the steering committee to have, so to speak, the "first and last word" on the various project processes. At the end of a process, the steering committee critically assesses the original plan and which parts of it have been implemented. Starting from the business case, the result must be compared to the requirements and if necessary – and this is common – the business case must be adjusted and re-negotiated. This enables risks to be identified in ongoing processes. These risks are then included in the risk log, and their impact on the business case must then be clearly identified. The "idea" behind the project: at this stage, the steering committee and management receive both an overview and a general sense of where they stand and what must be done during the next stage. This improves planning work for subsequent project stages.

To-dos at this stage:

- SB1: Planning a Stage
- SB2: Updating a Project Plan
- SB3: Updating a Project Business Case
- SB4: Updating the Risk Log
- SB5: Reporting Stage End

7. Closing a Project (CP)

PRINCE2 stresses the importance of a coordinated end to the project. This includes both completing the project and dismantling the (business) project structures, as well as analyzing the extent to which the project was completed according to plan, the deviations that occurred, and how these have impacted the rollout and the day-to-day workflow of the change introduced. Accordingly, the change in the standard operating process must be tested against the following criteria: is it effective and efficient? Does it offer added value? What impact has it had on the organization and what is the feedback from the workforce?

To-do list for this stage:

- CP1: Decommissioning a Project
- CP2: Identifying Follow-On Actions
- CP3: Evaluating a Project (see Beims et al. 2015)

Six Sigma

If a company has decided it wants to use Six Sigma as a process optimization method, it is possible to "trial" implementation of the method beforehand. At first, just a handful of senior managers receive training from consultants. During this training, the managers must utilize Six Sigma techniques to optimize a defined process in their company in order to qualify. This provides an efficient way of discovering whether the company can achieve added value by applying this method before committing to it. Since the method can be clearly quantified by calculating the return on investment, this clarifies whether profit due to optimization will exceed Six Sigma costs. This also enables successes to be distributed to all of the divisions within the company.

Six Sigma can be adjusted to various levels of management responsibility. Since heads of department, managers or knowledge workers perform distinct sets of tasks within the company, Six Sigma provides a matching "belt" for each role.

Many will ask themselves if Six Sigma is too complicated to apply within a company context. Yet the opposite is true, not least because Six Sigma offers a wide range of options for approaching a specific problem. For example, if turnaround time needs to be reduced from 24 hours to 12 hours for a specific product and production line, then Six Sigma is the right method to choose. Six Sigma becomes complex only if the process itself is complicated. If the company is already working to ISO or ITIL before optimization starts, this is beneficial, as standards will already be present in the processes and do not first need to be implemented. The process description is used here to achieve rapid, straightforward analysis and optimization. The first step is to examine whether potential for improvement is present, which is essentially a purely mathematical task. Optimization is then unproblematic and is implemented using the methods learned.

A second form of Six Sigma is also available, namely Lean Six Sigma. The principal difference is that Six Sigma improves single processes while Lean Six Sigma should be seen as a method for streamlining the entire company. Nor does Lean Six Sigma involve restructuring or strategic alterations: it simply applies familiar tools to streamline and optimize the company departments, processes and structures.

Since Six Sigma can be applied universally, it has no specific fields of application. Six Sigma can be used wherever there is potential to optimize processes. Sample applications include "identifying and resolving the causes of problem areas in the supply chain" (Gestmann 2008b) or "the use of statistical tools from Six Sigma to bring to light cause-effect relationships, and to identify key impact and success factors for ensuring successful placement [of recruitment agency personnel] [...]. When recruitment agencies want to improve the placement rate of their personnel, training is usually the first step. But if the recruitment agency's process organization and business processes can be optimally oriented to the company's goals as part of a Six Sigma project, then placement rates can be significantly increased. This has recently been substantiated by a Six Sigma project at a recruitment agency operating nationwide in Germany" (Business Wissen 2009). The methodology also helped to "identify supply chain problems and permanently remedy errors. [...] The mathematically-based approach applies statistical methods from Six Sigma to the key performance indicators used by the SCOR model. SCOR, short for Supply Chain Operation Reference, is a standardized model for describing business processes. 'This tool, which we call SCMAnalytics, enables data-driven statements to be made about the performance of the supply chain in question,' explains Michael Ferger from Six Sigma Germany. A total of 55 KPIs are measured, evaluated using the statistical methods from Six Sigma and then interpreted to arrive at a result. 'Each individual statement on the supply chain is qualified and substantiated with facts, figures and data,' Ferger continues. As a final step, the potential causes of problems already assigned to the KPIs are then analyzed and weighted. 'This enables performance obstacles to be immediately identified and removed,' adds Professor Schmieder, who was involved in tool development." (Gestmann 2008a). From the mid- to late 1990s, two Six Sigma waves reached Europe from the United States. Following its introduction at Motorola in 1987, Six Sigma successfully broke into the European market and the method is now applied in numerous US subsidiaries in Europe – including Kodak, Allied Signal and General Electric (see Töpfer 2007).

Annex to Chapter 7: Operational Quality: Zero Outage Ensures Reliability and Sustainability

A Sample Checklist from the Zero Outage Compliance Audits

INCIDENT MANAGEMENT

- A detailed Incident Report is created for all serious and critical incidents.
- Steps are taken to ensure that supplier tickets are opened before an incident is handed over to the RedPhone.
- Changes from the last few days are reviewed and all of the serious incidents are marked in the change list.
- Cross-checks/fire drills (simulations of faults) are planned and executed on a timely basis.
- In the event of a critical or potentially critical incident, RedPhone involvement is completed within 45 minutes.
- The Known Error Database is used for critical/high incidents.
- The customer business impact is verified before an incident is handed over to the RedPhone.
- Participation in the four-weekly Incident Management Community Meetings has been established for all time zones (Americas, EMEA/APAC).
- Participation in RedPhone and Global Problem Management early morning handover calls on all weekdays, for the purpose of reporting and handing over serious incidents previously handled in the service line, account, or local business unit.
- For serious or potentially critical incidents, the Yellow Local Phone is deployed within 20 minutes (local Lead Incident Manager or MoDs).

PROBLEM MANAGEMENT

- The standard (or customer-based) template for root cause analysis is utilized to the fullest extent.
- A company-wide Root Cause Database has been deployed and is also available in English for major incidents and early warnings.
- Problem resolution tracking (timetable and content) has been set up for major incidents, early warnings, on-demand incidents, and serious incidents.
- For the most important systems affected by incidents (e.g., Storage Boxes/SAN, Cloud/Appcom, Data Center Network, databases, middleware, operating systems), entries are created in the Known Error Database.
- Trend analyses are performed quarterly.
- Sign-off calls for serious incidents (high incidents) have been established (i.e., incidents not managed centrally by Global Problem Management).
- Actions for overdue root causes have been defined and implemented to reduce the backlog.
- A proactive Problem Management system has been set up. Routine incident analysis is carried out on a regular basis and appropriate actions have been defined to resolve the root causes permanently.
- For larger-scale incidents, early warnings and serious/special incidents, dedicated Lead Problem Managers (LPRMs) participate in "Get the day started" handover calls.
- Dedicated LPRMs participate in local/global Lessons Learned Sessions.
- Continual improvement work and quality initiatives are performed on the basis of root cause analyses.
- Fire drills (simulations of incidents under real-world conditions) are planned – if weaknesses are found in the alarm chain (internal/external).
- Actions for insights or weaknesses found in fire drills are defined and will be implemented in a timely manner.
- Critical landscapes and all involved service chains are regularly reviewed in order to keep information up-to-date at all times (date of last confirmed review by Service Delivery and Operations Manager no more than four weeks in the past).
- Root cause analyses include a written and confirmed root cause from Zero Outage-certified suppliers in the case of incidents that have been evaluated as "supplier errors".
- Actions for permanently resolving problems also include a "health check" provided by the supplier covering the latter's entire installed product and service base.
- Edits to entries in the Known Error Database are reviewed and approved by certified suppliers where appropriate.
- Problem tickets are proactively opened if the monthly SLA for a relevant customer is marked in RED.

CHANGE MANAGEMENT

- A Lead Change Manager (LCM) is given overall responsibility for each customer.
- LCM participation in global lessons learned sessions (monthly) is ensured.
- All major and significant changes have been implemented after approval from the Central Change Advisory Board (CCAB) (in the last three months).
- The local CAB has approved change models for recurring changes.
- Entries are made in the global Change Calendar on a regular basis.
- Relevant projects (category A, B) form part of the Change Calendar.
- High-risk changes and special-focus changes (e.g., "very high" risk, highly complex/critical changes) are announced at least 40 days before implementation, form part of the Change Calendar and are discussed by the CCAB and top management.
- All high-risk changes and special-focus changes (e.g., "very high" risk, very complex/critical changes) whose implementation was canceled were reviewed after completion.
- To facilitate plannability, the number of safeguarding calls required for high-risk and special-focus changes is no higher than three (incl. the last and final safeguarding call).
- The change grading of "Change implementation on time" (targets: 95% for all changes, 90% for "Major" and "Significant" changes) has been achieved.
- For all changes not implemented on time (60 minutes outside the change window), the reason for missing the deadline is investigated.
- The change grading of "Change implementation successful" (target: 98% for all change types) has been achieved.
- The cause of all non-successful changes is investigated.
- In the event of change-related incidents, the relevant CCAB/CAB members are involved to determine the cause (What went wrong during the change?).

CONFIGURATION MANAGEMENT

- A critical landscape has been described (for all critical service chains) and mapped out end-to-end in the CMDB.
- The correct criticality has been stored in the CMDB for each configuration item (and each service chain).
- All critical service chains have an SLA of at least xx.x%.
- Data quality is reviewed regularly with the aid of the standard KPIs and reports.

Glossary

3P: Umbrella term for the three most important factors for long-term quality assurance, namely: highly qualified personnel (**people**), simple, standardized processes (**processes**), and uniform, high performance platforms (**platforms**).

Agile Methodologies: Flexible approaches to project management, especially in software development. Agile models feature the rapid start of actual development work, frequent consultation with later users, constant testing, and the continuous improvement of the architecture.

Appliance: Design approach for a combined system of hardware and software optimized for this hardware. An appliance usually runs a single application or a small number of applications.

Capability Maturity Model Integration (CMMI): A model for the development of products and processes, whereby these are assigned to various capability levels that describe the respective degree of maturity for each.

Central Change Advisory Board (CCAB): A global de-escalation management unit that, as the approving authority, reviews all important and critical changes in IT back end systems, and monitors their implementation.

Change Management: An integral part of ITIL that describes a process which objective is to monitor and implement all adjustments to IT infrastructure efficiently, whilst minimizing operational risks to the provision of business services.

Claim Management: If deviations occur in the deliverables as agreed contractually, then measures are requested as part of contract change management, triggering the billing of additional expenses. A change request in this sense is the request from a partner due to a deviation.

Cloud Computing: IT infrastructure and applications (software, storage capacity, etc.) provided from a network, typically operated by a service provider. Data is no longer hosted on the company's own storage facilities, but in the provider's data center. In this way users are given dynamic, scalable IT resources that can be flexibly adjusted to changing needs. Billing is typically for the volume of services actually used.

Commodity Business: A sector characterized by increasingly homogeneous service portfolios.

Configuration Items (CIs): As defined by ITIL, any infrastructure component involved in business processes. Examples include: PCs, network devices, applications, servers and software.

Configuration Management Data Base (CMDB): Database for managing information about IT infrastructure and its configuration, used within configuration management. Helps companies to evaluate risks and impacts, and therefore reduce errors.

Continuous Improvement Program: Program that comprises the systematic, permanent and objective measurement of customer requirements on the one hand, as well as measures to ensure customer satisfaction on the other.

Critical Landscape: Overview of the customer's business-critical IT systems. Relevant for incident management, among other aspects (see also "Major Incidents").

Criticality: In IT, criticality expresses the significance of a system malfunction. Criticality is measured in levels: the higher the level, the more severe the expected impact in the event of a malfunction.

Customer Business Impact (CBI): Impact of a change or an incident on the customer's business processes.

De-Escalation Management: De-escalation management comprises the Central Change Advisory Board, global incident management and central problem management. De-escalation management is tasked with the rapid restoration of normal services after disruptions have occurred, and identifying the cause to initiate preventive actions.

End-to-End (E2E): In most cases, the provisioning of an IT service involves collaboration between various organizational units within the IT service provider's responsibility, as well as with external suppliers and partners. An end-to-end (E2E) perspective considers all stakeholders.

Failover Test: A failover is the unplanned switch from one technical component to another during a localized system failure. The failover test verifies proactively that the switch-over functions properly, thereby ensuring high availability.

Fire Drill: In IT, simulating system outages for the purpose of fault resolution training.

GxP Guidelines: The various ("x") guidelines for good working practice, especially in the fields of medicine, pharmacy and pharmaceutical chemistry. Examples: GMP (Good Manufacturing Practice) and GCP (Good Clinical Practice).

Health Check: At T-Systems, the routine checking of resources such as systems, applications or projects with the help of both standardized and specialized questionnaires.

ICT: Information and Communications Technology. Describes the combination of information technology (IT) with (tele)communications (C).

Incident Management: An integral part of ITIL that describes a process designed to resolve faults (incidents) in IT operations.

Insourcing: Having services performed by (own or third-party) personnel working at the same site (see also onshoring).

Intensive Care: At T-Systems, a standardized approach to analysis and improvement for resolving quality problems at major customers.

ISAE 3402: Abbreviation for the International Standard on Assurance Engagements. Certifies the control system at a service provider. ISAE 3402 has superseded the US standard audit report SAS-70 and serves as the basis for an integrated control system.

IT Infrastructure Library (ITIL): Framework of best practices for IT processes. ITIL was first published in 1989 by the UK Office of Government Commerce (OGC) and has since undergone several significant revisions.

Key Performance Indicator (KPI): A metric used to assess the achievement of defined target values.

Lean Management: A systematic method to optimize or streamline processes within a manufacturing system.

Major Incident (MI): A severe fault affecting an IT system, or its complete failure. An MI disrupts a key business process in a company in such a way that significant damage (loss of reputation or financial loss) is expected as a consequence.

Manager on Duty (MoD): An IT service provider staff member with a management role in the operations unit(s), who is notified in the event of escalations and emergencies. Primary tasks are coordinating crisis meetings, deciding on resource usage, providing support for escalation to third parties, and organizing the confirmation chain for emergency changes.

Mean Time to Repair (MTTR): Average time required to restore IT systems following a system outage.

Nearshoring: Outsourcing services to neighboring countries (see also offshoring, onshoring).

Offshoring: Outsourcing services to other (distant) countries (see also nearshoring, onshoring).

Onshoring: Outsourcing services to providers in the same country (see also nearshoring, offshoring).

Organizational Structure: A company's hierarchical "backbone." The organizational structure describes which tasks are completed by which resources (people and work equipment). Each company also possesses a process structure, which represents the process flows.

Outsourcing: Moving of services or units to external providers.

Problem Management: An integral part of ITIL that describes a process designed to identify the causes of faults (incidents) and prevent their future reoccurrence. Results may either be "known errors" or workarounds: Details of these are provided to incident management to prevent a repeat occurrence of the fault.

Quality Academy: At T-Systems, a unit that uses a modular, self-study model based on individual job descriptions to provide group wide training and a uniform certification system. The training content and certification addresses various quality topics with the aim of ensuring employees "live and breathe" quality.

Quality Roadmap: A unique method within Zero Outage, used to manage all quality-relevant risks at T-Systems according to a clearly defined structure. To this end, around 280 identified individual risks are grouped into 40 categories. The maps in the Quality Roadmap derived in this way define initiatives and measures designed to permanently eliminate these risks. The three most important components here are quality-certified employees, standardized processes, and modern, high availability platforms.

Root Cause: Primary cause of an incident (fault).

Root Cause Rate/Root Cause Rate in Time: Key performance indicator that tracks the speed and timeliness with which the primary causes of incidents are being detected.

Service Delivery Manager: Customer point of contact. Responsible for customer requirements, the contract, financial planning, operational escalations, regular reviews and SLA reporting.

Service Improvement Program (SIP): Defines measures to improve processes and services within an agreed period of time, along with measurable key figures for progress and results. Can build on the results of a service review, for example, and has the goal of closing the gaps that the review has identified.

Service Level Agreement (SLA): Agreement between the customer and the provider, in which the quality of a service is documented using metrics such as bandwidth, availability, etc.

Standardization: In IT, the act of using empirical values to make processes, products and services more uniform in order to design more efficient processes (from a cost and productivity perspective) and to offer services at a maximum level of quality and cost-effectiveness.

Supply Chain: Represents the entire chain of added value and resources from individual parts/services to the final product (or service).

WAN: A Wide Area Network (WAN) is a computer network extending over countries and continents to which an unlimited number of computers can be connected.

Zero Outage: An integrated quality program at T-Systems consisting of standards in the fields of people, processes and platforms (see 3Ps), with the aim of securing customer project reliability by ensuring IT services are operated with a minimum of errors.

References

Beims, Martin (2012): IT-Service Management in der Praxis mit ITIL (3rd edition). Munich: Carl Hanser Verlag, p. 225f.

Beims, Martin; Ziegenbein, Michael (2015): IT-Service Management in der Praxis mit ITIL (4th edition). Munich: Carl Hanser Verlag Munich, p. 310–325.

Bitkom (2016): Digitalisierung der Wirtschaft nimmt Fahrt auf. https://www.bitkom.org/Presse/Presseinformation/Digitalisierung-der-Wirtschaft-nimmt-Fahrt-auf.html. Accessed: 06/27/2016.

Bon, Jan van; Jong, Arjen de; Kolthof, Axel; Pieper, Mike; Tjassing, Ruby; van der Veen, Annelies; Verheijen, Tienke (2010): ITIL V – Das Taschenbuch. Zaltbommel: Van Haren Publishing, p. 25–64.

Business Wissen (2009): Six-Sigma-Projekt schafft Grundlage für höhere Vermittlungserfolge. http://www.business-wissen.de/artikel/personalvermittlung-six-sigma-projekt-schafft-grundlage-fuer-hoehere-vermittlungserfolge/. Accessed: 06/14/2016.

Capgemini (2015): Studie IT-Trends 2015 – Digitalisierung gibt Zusammenarbeit zwischen Business und IT eine neue Qualität. https://www.de.capgemini.com/resource-file-access/resource/pdf/it-trends-studie-2015.pdf. Accessed: 06/09/2016.

DeMarco, Tom; Lister, Tim (2003): Bärentango, Mit Risikomanagement Projekte zum Erfolg führen. Carl Hanser Verlag GmbH & Co. KG.

Ebel, Nadin (2011): PRINCE2:2009 – für Projektmanagement mit Methode. Munich: Addison-Wesley, p. 103–111.

Fröhlich, Martin; Glasner, Kurt (2007): IT-Governance. Wiesbaden: Betriebswirtschaftlicher Verlag Dr. Th. Gabler; GWV Fachverlage GmbH, p. 6.

Gestmann, Michael (2008a): Auf Fehlersuche in der Lieferkette. In: Industrieanzeiger Online. http://www.industrieanzeiger.de/home/-/article/12503/11824006/Auf+Fehlersuche+in+der+Lieferkette/art_co_INSTANCE_0000/. Accessed: 06/14/2016.

Gestmann, Michael (2008b): Six Sigma: Probleme in der Lieferkette erkennen und abstellen. http://www.business-wissen.de/artikel/six-sigma-probleme-in-der-lieferkette-erkennen-und-abstellen/. Accessed: 06/28/2016.

Glaap, Winfried (1993): ISO 9000 leichtgemacht. Munich, Vienna: Carl Hanser Verlag Munich Vienna.

Gorecki, Pawel; Pautsch, Peter R. (2014): Praxisbuch Lean Management – Der Weg zur operativen Excellence (2nd edition). Munich: Carl Hanser Verlag Munich, p. 3–27.

Görgen, Peter (2015): Das Problemkind „Problem Management". In: Materna Monitor (2015) no. 2, p. 33–35.

International Organization for Standardization (2015): Quality management principles. http://www.iso.org/iso/pub100080.pdf. Accessed: 06/21/2016.

ISG (2015): ISG Study. https://www.t-systems.com/de/en/about-t-systems/company/newsroom/news/news/isg-study-validates-t-systems-strategy-225844. Accessed: 07/04/2016.

ITSM Group (2015): Wachsender Fokus auf die IT-Servicequalität. https://www.itsm-consulting.de/news-events/news-feed/wachsender-fokus-auf-die-it-servicequalitaet. Accessed: 06/06/2016.

Itwnet (2016): ISO 9001:2000 and ISO/IEC 20000:2005 Comparison. http://www.itwnet.com/system/files/ISO9001_20000_Cross-reference.pdf. Accessed: 06/28/2016.

Kaspersky (2013): Security Bulletin 2013 / 2014. http://media.kaspersky.com/de/business-security/Kaspersky Security Bulletin 2013_2014_ebook_deutsch.pdf. Accessed: 06/10/2016.

Kasulke, Stephan (2013a): Maßnahmen zur kurzfristigen Qualitätsverbesserung. In: Abolhassan, Ferri (ed.): The Road to a Modern IT Factory Industrialization – Automation – Optimization Wiesbaden: Springer Gabler, p. 71–78.

Kasulke, Stephan (2013b): Maßnahmen zur mittel- und langfristigen Qualitätsverbesserung. In: Abolhassan, Ferri (ed.): The Road to a Modern IT Factory. Wiesbaden: Springer Gabler, p. 109–113.

Kasulke, Stephan (2014): Continuous Improvement – Qualität optimieren und Kundenzufriedenheit garantieren. In: Abolhassan, Ferri (ed.): Kundenzufriedenheit mit IT-Outsourcing. Das Optimum realisieren. Wiesbaden: Springer Gabler, p. 41–52.

Kayenta (2015): PMP Zertifizierung – Project Management Professional PMP nach PMI. http://web.archive.org/web/20151014233229/http://www.kayenta.de/seminar-training/projektmanagement/pmp-zertifizierung-project-management-professional-pmp-nach-pmi.html. Accessed: 06/14/2016.

PMI (2014): Project Management Institute. http://h10076.www1.hpe.com/education/standards_faq_brand.pdf. Accessed: 06/14/2016.

PwC (2012): IT-Sourcing-Studie 2012 – Aktuelle IT-Sourcing-Perspektiven erkennen und nutzen. http://www.pwc.at/presse/2012/pdf/studie-it-sourcing-2012.pdf. Accessed: 06/09/2016.

PwC (2015): IT-Sourcing-Studie – Die Perspektive der Anbieter. http://www.conect.at/uploads/tx_posseminar/20150319_ITSM_PwC_IT_Sourcing_v1.0_2_.pdf. Accessed: 06/09/2016.

Schiefer, Helmut; Schitterer, Erik (2008): Prozesse optimieren mit ITIL (2nd edition). Wiesbaden: GWV Fachverlage GmbH, p. 12.

SixSigma (2015): http://www.six-sigma.de/

Stych, Christof; Zeppenfeld, Klaus (2008): ITIL. Berlin: Springer-Verlag Berlin Heidelberg, p. 15.

T-Online (2014): Hacker durch Zufall: Jugendliche knacken Bankautomaten. http://www.t-online.de/computer/sicherheit/id_69792478/14-jaehrige-knacken-bankautomaten-mit-handbuch-aus-dem-internet.html. Accessed: 06/10/2016.

Töpfer, Armin (2007): Six Sigma: Konzeption und Erfolgsbeispiele für praktizierte Null-Fehler-Qualität (4th edition). Springer Berlin, Heidelberg, New York.

TÜV (2015): ISO 9001 – Qualität mit System. http://www.tuev-sued.de/management-systeme/iso-9001. Accessed: 06/14/2016.

TÜV NORD AUSTRIA (2016): Informationen – ein kostbares Gut. https://www.tuv-nord.com/at/de/informations-technologie/iso-27001-616.htm. Accessed: 06/14/2016.

TÜV NORD GROUP (2014): TÜV NORD CERT – Zertifizierung von Qualitätsmanagementsystemen nach DIN EN ISO 9001. https://www.tuev-nord.de/fileadmin/Content/Global/TUEV_NORD_Archiv/pdf/pdb-iso-9001.pdf. Accessed: 06/14/2016.

ZDNet (2015): Apple: iTunes Store, App Store outage cased by 'internal' error. http://www.zdnet.com/article/apples-itunes-store-app-store-experiencing-outages/. Accessed: 06/27/2016.

Thank you

We would like to thank everyone involved for all of their hard work.

Special thanks go to the following people for helping to make this book a reality and for supporting the authors:

 **Heike Bayerl** has been Vice President for Security Compliance and Quality Management at T-Systems International GmbH since 2012. In this position she is responsible for the Quality Academy worldwide. She also plays a key role in the management and auditing of issues relating to the IT operations of the internal Security and Quality Management department. Heike Bayerl has a degree in communications engineering from Offenburg University of Applied Sciences (university for engineering, business and media). She is also a certified Six Sigma Black Belt.

 **Dr. Jürgen Herczeg** has been Vice President Process & Quality Management at T-Systems International GmbH since 2013. He has worked in software project development, consulting, quality assurance and management for 20 years. Since the end of 2005, he has been part of a team of software architects who are drawing up company-wide standards and development programs for Software Engineering & Architecture at T-Systems. Dr. Jürgen Herczeg studied computer science at the University of Stuttgart and earned a doctorate in 1995.

Ines-Maria Böckl has been working as a project manager at T-Systems International GmbH since 2006, where she manages the projects of various organizations and European subsidiaries. She oversaw the establishment of strategic production sites and contributed to the creation of a hardware and capacity management system. She also designed the resource management system for three European subsidiaries. In 2016, she moved to the Quality Line Office at T-Systems. Even while studying for an undergraduate degree in religious pedagogy, her focus was on communication and adult education. She later earned an MBA with a focus on intercultural learning, work and management from the University of Louisville.

Nadine Schmidt has been a Communications Expert at T-Systems International GmbH since 2013, where she develops internal and external communication concepts relating to Zero Outage and ensures that they are implemented appropriately for each target group. She started her career in the Group in 2000 after studying international business studies at Dortmund University of Applied Sciences and the University of Abertay Dundee in Scotland. She has a business administration degree and a Bachelor of Arts with first-class honors. She additionally qualified as a PR consultant in 2008.

Christian Braunsteiner has been involved in the ICT industry for over 15 years. After completing commercial training with a focus on business information systems, he joined the Service and System Operations departments of mobile communications provider max.mobil (today T-Mobile Austria) in 2000. He has worked in the Quality department of T-Systems International GmbH since 2008. Christian Braunsteiner earned a bachelor's degree from the International Management Center in Krems and a master's degree from Danube University Krems.

We would also like to thank the following people who made valuable contributions to the content and coordination of this book: Doris Reitter, Björn Petersen, Sascha Lisson, Joachim Bausch, Falk Reckert and Bernd Najewitz. And last but not least, thank you to PSM&W (particularly Birgit Wölker and Dominique-Silvia Kemp). Specials thanks go to Fergal O'Donnell and Karel van Zyl who supported us with their native language skills and professional background in producing this book in English.

Zeitfracht Medien GmbH
Ferdinand-Jühlke-Straße 7
99095 Erfurt, Deutschland
produktsicherheit@kolibri360.de